Saas-Fee Advanced Course 48

More information about this series at http://www.springer.com/series/4284

Tiziana Di Matteo · Andrew King ·
Neil J. Cornish

Black Hole Formation and Growth

Saas-Fee Advanced Course 48

Swiss Society for Astrophysics and Astronomy
Edited by Roland Walter, Philippe Jetzer,
Lucio Mayer and Nicolas Produit

Springer

Authors
Tiziana Di Matteo
McWilliams Center for Cosmology
Carnegie Mellon University
Pittsburgh, USA

Andrew King
Department of Physics and Astronomy
University of Leicester
Leicester, UK

Neil J. Cornish
Department of Physics
Montana State University
Bozeman, USA

Volume Editors
Roland Walter
Department of Astronomy
University of Geneva
Geneva, Switzerland

Philippe Jetzer
Physik-Institut
University of Zurich
Zürich, Switzerland

Lucio Mayer
Institute for Computational Science
University of Zurich
Zürich, Switzerland

Nicolas Produit
Department of Astronomy
University of Geneva
Geneva, Switzerland

This Series is edited on behalf of the Swiss Society for Astrophysics and Astronomy: Société Suisse d'Astrophysique et d'Astronomie, Observatoire de Genève, ch. des Maillettes 51, CH-1290 Sauverny, Switzerland.

ISSN 1861-7980 ISSN 1861-8227 (electronic)
Saas-Fee Advanced Course
ISBN 978-3-662-59798-9 ISBN 978-3-662-59799-6 (eBook)
https://doi.org/10.1007/978-3-662-59799-6

Cover illustration: Scientists have obtained the first image of a black hole, using Event Horizon Telescope observations of the center of the galaxy M87. The image shows a bright ring formed as light bends in the intense gravity around a black hole that is 6.5 billion times more massive than the Sun. Credit: Event Horizon Telescope Collaboration

This Springer imprint is published by the registered company Springer-Verlag GmbH, DE part of Springer Nature.
The registered company address is: Heidelberger Platz 3, 14197 Berlin, Germany

Preface

The 48th "Saas-Fee Advanced Course" of the Swiss Society for Astrophysics and Astronomy (SSAA) was held from 28 January to 3 February 2018 in Saas-Fee, in the Swiss Alps. It was very timely devoted to:

Black Hole Formation and Growth

and attended by 119 participants. The Saas-Fee courses are intended mainly for postgraduate, Ph.D. students, astronomers and physicists who wish to broaden their knowledge. The lectures were organised in the morning and late afternoon leaving free time for informal discussions, studies and outdoor activities in the afternoons.

This advanced course provided three comprehensive and up-to-date reviews covering the gravitational wave breakthrough, our understanding of accretion and feedback in supermassive black hole and the relevance of black hole to the Universe structure since the Big Bang. The lectures were given by three world experts in the field:

Prof. Tiziana Di Matteo (Carnegie Mellon University, USA)
Tiziana Di Matteo is a Professor in the McWilliams Center for Cosmology of the Physics Department at Carnegie Mellon University, USA. She received her Ph.D. in 1998 at Cambridge University, UK. She was a Chandra Fellow at Harvard and a junior faculty member at the Max Planck Institute for Astrophysics in Germany. She is a theorist with expertise in both high energy astrophysics and cosmology. Her interests focus on state-of-the-art cosmological simulations of galaxy formation with special emphasis on modelling the impact of black holes on structure formation in the Universe.

Prof. Andrew King (University of Leicester, UK)
Andrew King is Professor of Theoretical Astrophysics at the University of Leicester and holds visiting appointments at the Universities of Amsterdam and Leiden. During his career, he has been awarded a PPARC Senior Fellowship, the Gauss

Professorship at the University of Goettingen, a Royal Society Wolfson Merit Award, and the RAS Eddington Medal. He is an author and co-author of several books, including Stars, a Very Short Introduction, and Accretion Power in Astrophysics. His research interests include accretion disc structure, supermassive black hole growth and feedback, active galactic nuclei, compact binary evolution, and ultraluminous X-ray sources.

Prof. Neil J. Cornish (Montana State University, USA)
Neil J. Cornish is Regents Professor of Physics and Director of the eXtreme Gravity Institute at Montana State University. He completed his Ph.D. at the University of Toronto, followed by postdoctoral fellowships in Steven Hawking's group at the University of Cambridge and in David Spergel's group at Princeton University. He is a multiwavelength gravitational-wave astronomer, and he is a member of the Laser Interferometer Gravitational-Wave Observatory (LIGO) Scientific Collaboration, the North American Nanohertz Observatory for Gravitational Waves (NANOGrav) Collaboration and the NASA Laser Interferometer Space Antenna (LISA) Science Study team.

Exactly 100 years before this Saas-Fee course, on 31 January 1918, Einstein submitted his paper entitled "Gravitational waves" to be presented at the Prussian Academy meeting held on 14 February. Actually, Einstein started to think on gravitational waves (at least as far it is documented) in 1913 when he was still Professor at ETH in Zürich. At the 85th Congress of the German Natural Scientists and Physicists (9 September 1913) in Vienna, Max Born asked Einstein about the speed of propagation of gravitation, in particular, whether it would be that of light. Einstein replied that it is extremely simple to write down the equations for the case in which the disturbance in the field is extremely small and this is what Einstein then did in his 1916 paper "Approximate Integration of the Field Equations of Gravitation", with some mistake, and in a more correct way in 1918.

It took a century from Einstein's papers to the actual detection of a gravitational wave (and the ultimate proof that black holes exist). A pretty important step, demonstrating that the content of this book is closer to reality than it would have been just a few years ago.

We are very grateful to the lecturers for their enthusiasm in communicating their deep knowledge, their brilliant lectures, as well as for writing the rich manuscripts composing this book. We extend our warmest thanks to the course secretaries, Martine Logossou and Marie-Claude Dunand, for their effective administration and organisational help during the course. We also would like to thank our students and collaborators helping in Saas-Fee and finalising the manuscripts, in particular, V. Sliusar, C. Panagiotou, M. Balbo, E. Lyard, T. Bernasconi and M. Kole.

Saas-Fee provided a very pleasant environment with two metres of fresh snow and an entirely sunny week. We enjoyed the first century birthday of gravitational wave on 31 January 2018, with a concert of the Swiss ethno-electronic music band "Vouipe", who composed the track "Black Hole in Saas-Fee"[1] based on the

[1]Available at http://vouipe.com/.

space-time chirp of GW150914! We also enjoyed a concert from Moncef Genoud and Ernie Odoom. These were magical evenings, and we would like to thank again the outstanding performers for their delighting music.

Finally, this course would not have been possible without the financial support of the Swiss Society for Astrophysics and Astronomy, the Société Académique de Genève and the Universities of Geneva and Zürich. We are very grateful to these organisations for their contributions, which allowed the participants to attend a very interesting and successful course.

The course organisers

Geneva, Switzerland — Roland Walter, Nicolas Produit
Zürich, Switzerland — Philippe Jetzer, Lucio Mayer

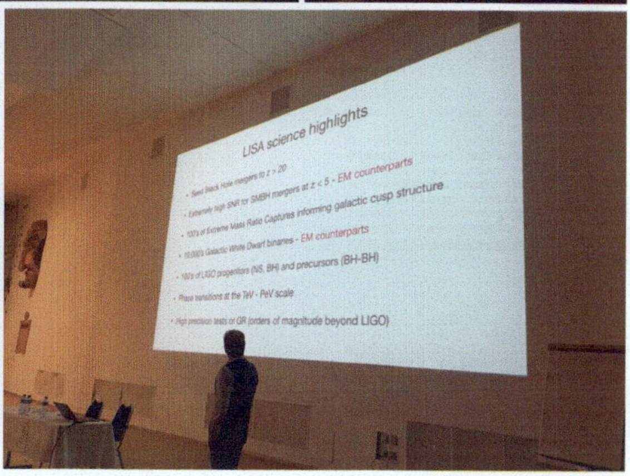
LISA science highlights
Extremely high SNR for SMBH mergers at z < 5 - EM counterparts
100's of Extreme Mass Ratio Captures informing galactic cusp structure
Galactic White Dwarf binaries - EM counterparts
100's of LIGO progenitors (NS, BH) and precursors (BH-BH)
Phase transitions at the TeV - PeV scale
High precision tests of GR (orders of magnitude beyond LIGO)

Contents

Black Hole Merging and Gravitational Waves

Neil J. Cornish

Contents

N. J. Cornish (✉)
Department of Physics, Montana State University, Bozeman, MT 59717, USA
e-mail: ncornish@montana.edu
URL: http://www.montana.edu/xgi/

R. Walter et al. (eds.), *Black Hole Formation and Growth*, Saas-Fee Advanced Course 48,
https://doi.org/10.1007/978-3-662-59799-6_1

Abstract I was tasked with covering a wide swath of gravitational wave astronomy—including theory, observation, and data analysis—and to describe the detection techniques used to span the gravitational wave spectrum—pulsar timing, ground based interferometers and their future space based counterparts. For good measure, I was also asked to include an introduction to general relativity and black holes. Distilling all this material into nine lectures was quite a challenge. The end result is a highly condensed set of lecture notes that can be consumed in a few hours, but may take weeks to digest.

1 Introduction

In writing up these lecture notes I have mostly followed the order in which the material was presented in Saas Fee, with the exception of the discussion of the detectors, which have been grouped here in a single section. My goal is not to write a textbook on each topic—many excellent texts and review articles on general relativity and gravitational wave astronomy already exist (see e.g. [11, 16, 37, 40, 50]). Rather, I try to highlight the key concepts and techniques that underpin each topic. I also strive to provide a unified picture that emphasizes the similarities between pulsar timing, ground base detectors and space based detectors, and commonalities in how the data is analyzed across the spectrum and across source types.

2 General Relativity

The historical course that lead Einstein to develop the general theory of relativity had many twists and turns, but as Einstein reflected in 1922, one of his primary goals was to understand the equivalence between inertial mass and gravitational mass "It was most unsatisfactory to me that, although the relation between inertia and energy is so beautifully derived [in Special Relativity], there is no relation between inertia and weight. I suspected that this relationship was inexplicable by means of Special Relativity" [43]. Einstein found the resolution to this conundrum by adopting a geometrical picture that generalized Minkowski's description of special relativity to allow for spacetime curvature.

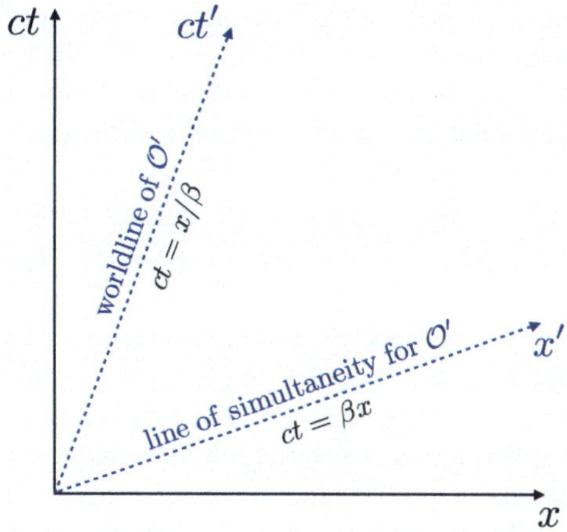

Fig. 1 A spacetime diagram shown in the the rest frame of observer $\mathcal{O}$. Observer $\mathcal{O}'$ is moving at velocity v with respect to $\mathcal{O}$ in the x direction

2.1 Special Relativity

Minkowski showed that phenomena such as time dilation and length contraction followed naturally as a consequence of space and time being combined into a single spacetime geometry with distances measured by the invariant interval

$$ds^2 = -c^2dt^2 + dx^2 + dy^2 + dz^2\,. \tag{1}$$

The $t =$ const. spatial section of this geometry is ordinary three dimensional Euclidean space, which is invariant under translations and rotations, and can be described by the special Euclidean group E(3) = SO(3) × T(3). The full Minkowski spacetime is invariant under the Poincaré group, which is made up of translations in time and space $T(1, 3)$, rotations in time and space SO(1, 3), otherwise known as the Lorentz group. The Lorentz group includes regular spatial rotations SO(3), and boosts, which can be thought of as a hyperbolic rotation in a plane that includes a time-like direction.

The key results of special relativity can be derived by considering motion that is restricted to the $(1 + 1)$ dimensional sub-manifold spanned by coordinates (t, x) with invariant interval $ds^2 = -c^2dt^2 + dx^2$. Rotations in this two-dimensional Minkowski space leave fixed hyperbolae, $x^2 - c^2t^2 = \pm a^2$, just as rotations in two dimensional Euclidean space leave fixed circles, $x^2 + y^2 = a^2$. The coordinates (t, x) define a frame of reference $\mathcal{O}$. An observer at rest in this coordinate system will follow the trajectory $x =$ const.: in other words, a line parallel to the t-axis. A particle moving at velocity v in the positive x-direction will follow the trajectory (worldline) $x = vt +$ const. We can perform a boost to a new reference frame $\mathcal{O}'$, with coordinates (t', x'), where the particle is a rest: $x' =$ const. This implies that the two coordinate systems are related: $x' = \gamma(x - vt)$, where γ is a constant. Objects at rest in frame $\mathcal{O}$ will be moving at velocity $-v$ in the x' direction in frame $\mathcal{O}'$, so it follows that $x = \gamma(x' + vt')$. Solving for t' we find $t' = \gamma(t + (1-\gamma^2)/(v\gamma^2)x)$. Invariance of the interval $x'^2 - c^2t'^2 = x^2 - c^2t^2$ fixes the constant to be $\gamma = 1/\sqrt{1-\beta^2}$, where $\beta = v/c$. Thus we have derived the following coordinate transformation for a boost with velocity v in the positive x direction:

$$\begin{aligned} ct' &= \gamma(ct - \beta x) \\ x' &= \gamma(x - vt) \,. \end{aligned} \tag{2}$$

The transformation can be viewed as a hyperbolic rotation with $\cosh\eta = \gamma$ and $\sinh\eta = \beta\gamma$, were the "angle" $\eta = \text{arctanh}\,\beta$ is called the rapidity. Lines of simultaneity in $\mathcal{O}'$ have $t' = 0$, and lie parallel to the line $x = ct/\beta$ in frame $\mathcal{O}$. Figure 1 displays a spacetime diagram illustrating the relationship between the two reference frames.

Classic results such as time dilation and length contraction follow directly from the spacetime geometry of Minkowski space. For example, consider two events A, B along the worldline of observer $\mathcal{O}'$. The proper time elapsed as measured by a clock carried by observer $\mathcal{O}'$ is $T' = \Delta t'$, while the time elapsed as measured by a clock carried by observer $\mathcal{O}$ is $T = \Delta t$. The invariant interval is

$$\Delta s_{\rm AB} = -(c\Delta t')^2 = -(c\Delta t)^2 + \Delta x^2 \,. \tag{3}$$

Since $\Delta x = v\Delta t$ we find that

$$T = \gamma T' \,, \tag{4}$$

so that the time elapsed in the static frame is greater than the time elapsed in the moving frame. Next consider a rod of proper length L' moving at velocity v relative to observer $\mathcal{O}$. At a given instant, the ends of the rod at at events D, F in frame $\mathcal{O}$ and at events D, E in frame $\mathcal{O}'$. Thus the length of the rod in frame $\mathcal{O}$ is $L = \Delta s_{\rm DF}$, while the length of the rod in frame $\mathcal{O}$ is $L' = \Delta s_{\rm DE}$. Using the invariance of the interval we have

$$\begin{aligned} \Delta s_{\rm DF}^2 &= L^2 = \Delta x^2 = \Delta x'^2 - (c\Delta t')^2 \\ \Delta s_{\rm DF}^2 &= L'^2 = \Delta x'^2 = (\Delta x + v\Delta t)^2 - (c\Delta t)^2 \,. \end{aligned} \tag{5}$$

Incorporating the time dilation found earlier, $\Delta t = \gamma\Delta t'$, we find the lengths to be related:

$$L = L'/\gamma \,, \tag{6}$$

so that the rod appears shorter in the static frame. The spacetime diagram makes it clear that this discrepancy is due to the two frames having different lines of simultaneity (Fig. 2).

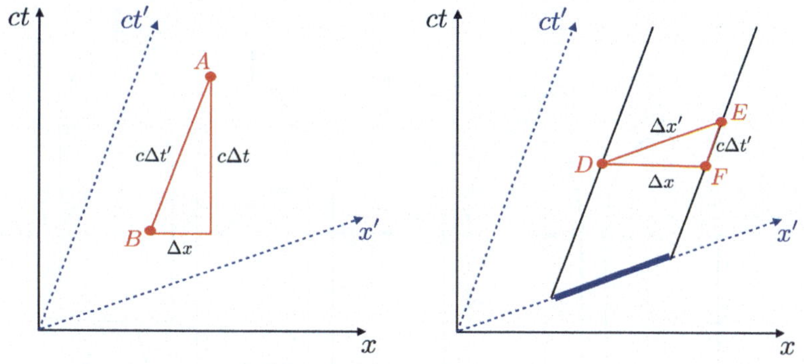

Fig. 2 Spacetime diagrams illustrating time dilation (left) and length contraction (right)

2.2 *The Equivalence Principle*

Einstein, like Newton before him, was struck by the equivalence of the inertial mass that appears in the relation between force and acceleration $\mathbf{F} = m_I \mathbf{a}$ and the gravitational charge, or mass, that appears in Newton's gravitational force law $\mathbf{F}_G = -Gm_G M\mathbf{r}/r^3$. He also noted that inertial frames of reference play a special role in both Newtonian mechanics and special relativity. In Newtonian mechanics, objects in a non-inertial frame that is uniformly accelerating and rotating experience "pseudo-forces" of the form

$$\mathbf{F}_\mathrm{P} = -m_I \mathbf{a} - 2m_I \boldsymbol{\omega} \times \mathbf{v} - m_I \boldsymbol{\omega} \times (\boldsymbol{\omega} \times \mathbf{x})\,. \tag{7}$$

Since these forces are a coordinate effect, they must scale with the inertial mass. The first term in the above expression is referred to as a rectilinear force, the second term is called the Coriolis force, while the third term is called the centrifugal force. The "pseudo" moniker is perhaps a little misplaced—hurricanes that get stirred up by the Coriolis force due to the rotation of the Earth are real enough. Einstein suggested that the rectilinear term be identified with a uniform gravitational field. "I was sitting in a chair in the patent office in Bern when all of a sudden a thought occurred to me: if a person falls freely he will not feel his own weight. I was startled. This simple thought made a deep impression upon me. It impelled me towards a theory of gravitation" [43]. In other words, the acceleration $g \simeq 9.8\,\mathrm{m\,s^{-2}}$ that we experience while sitting or standing on the surface of the Earth is due to us being in a non-inertial frame of reference. Take away the ground, say by jumping into a mineshaft, and the "force of gravity" goes away. The equivalence of inertial and gravitational mass follows naturally from the equivalence of uniform gravitational fields and uniform accelerations.

Einstein set out to incorporate this insight into a modification of special relativity that could account for gravitational effects. The connection to coordinate transformations suggested a geometrical approach, which caused Einstein to pay more at-

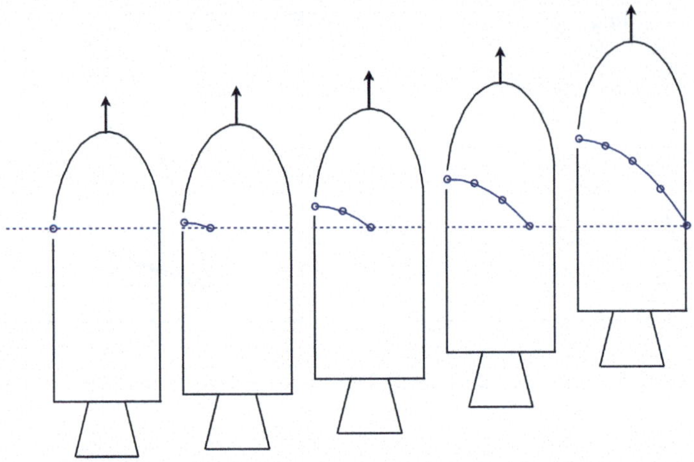

Fig. 3 Path of light as seen in a uniformly accelerated reference frame. In an inertial frame the light follows the straight-line dotted path, while in the accelerated frame of the rocket the path appears to follow a curved path shown here as a solid line

tention to Minkowski's geometrical formulation of special relativity. Einstein began by showing that the path of light seen by a uniformly accelerated observer could be interpreted in terms of spacetime geometry where the speed of light depends on position.

Consider the picture in Fig. 3 where a rocket accelerates from rest with uniform acceleration a in the positive z direction, and a photon traveling in the positive x direction enters a window in the rocket at time t. The photon follows the path $x = ct, z = 0$, shown as a horizontal dotted line, in the inertial frame where the rocket was originally at rest. A first the velocity of the rocket $v_z = at$ is much less than the speed of light, and the coordinates in the two frames are related such that $t' = t$, $x' = x$ and $z' = z - \frac{1}{2}at^2$. Thus, in the non-inertial frame of the rocket, the photon follows the parabola $z' = -ax'^2/(2c^2)$. Einstein showed that this "bending of light" could be derived from the line element for a uniformly accelerated observer, which to leading order in a has the form [18]

$$ds^2 = -c^2\left(1 + \frac{2az'}{c^2}\right)dt'^2 + dx'^2 + dy'^2 + dz'^2 . \tag{8}$$

To confirm that light paths in this geometry are indeed parabolas, we need to derive the geodesic equation, which describes the shortest/straightest paths in spacetime.

Using the short-hand notation $\mathbf{x} \to \{x^\mu\} = \{ct, x, y, z\}$ and $ds^2 = g_{\mu\nu}(\mathbf{x})\, dx^\mu dx^\nu$, with summation implied on repeated indices, the path length between events A, B, is given by

$$S = \int_A^B \sqrt{-ds^2} = \int_{\lambda_A}^{\lambda_B} \left(-g_{\mu\nu}\frac{dx^\mu}{d\lambda}\frac{dx^\nu}{d\lambda}\right)^{1/2} d\lambda \equiv \int_{\lambda_A}^{\lambda_B} L\left(x^\mu, \frac{dx^\mu}{d\lambda}\right) d\lambda. \quad (9)$$

Holding the end points fixed and extremizing the path length yields the Euler–Lagrange equations

$$\frac{d}{d\lambda}\left(\frac{\partial L}{\partial(dx^\alpha/d\lambda)}\right) = \frac{\partial L}{\partial x^\alpha} \quad (10)$$

which evaluate to

$$\frac{d^2x^\alpha}{d\lambda^2} = -\frac{1}{2}g^{\alpha\beta}(g_{\beta\mu,\nu} + g_{\beta\nu,\mu} - g_{\mu\nu,\beta})\frac{dx^\mu}{d\lambda}\frac{dx^\nu}{d\lambda} \equiv -\Gamma^\alpha_{\mu\nu}\frac{dx^\nu}{d\lambda}\frac{dx^\nu}{d\lambda}. \quad (11)$$

Here commas denote partial derivatives $h_{,\mu} = \partial h/\partial x^\mu$, and the collection of metric derivatives appearing on the right-hand side of the above equation are referred to as the Christoffel symbol $\Gamma^\alpha_{\mu\nu}$. Note that $g^{\alpha\beta}$ denotes the components of the inverse metric tensor, so that $g^{\alpha\beta}g_{\alpha\kappa} = \delta^\alpha_\kappa$. The geodesic equation can be simplified by introducing the notation $u^\alpha = dx^\alpha/d\lambda$ for the 4-velocity and $\nabla_\beta u^\alpha = u^\alpha_{\ ,\beta} + u^\nu\,\Gamma^\alpha_{\beta\nu}$ for the covariant derivative. With these definitions we have $d/d\lambda = u^\alpha\nabla_\alpha$, and the geodesic equation takes the more compact form

$$u^\beta\nabla_\beta u^\alpha = 0. \quad (12)$$

Returning to the metric for a uniformly accelerated observer, we find $\Gamma^z_{tt} = a$ and all others zero. For a photon with initial velocity $\mathbf{U} \to (1, c, 0, 0)$ the geodesic equations integrate to give $t' = \lambda$, $x' = ct'$ and $z' = -\frac{1}{2}at'^2$. This confirms that photons do indeed follow parabolic paths in the $x' - z'$ spacetime with line element given in Eq. (8). The result can be generalized to describe uniform acceleration in any direction and uniform rotation about any axis. The line element for this non-inertial frame is given by (dropping the primes to simplify the notation)

$$ds^2 = -((c + \mathbf{a}\cdot\mathbf{x})^2 - (\boldsymbol{\omega}\times\mathbf{x})^2)dt^2 + 2c(\boldsymbol{\omega}\times\mathbf{x})_i dx^i dt + dx^2 + dy^2 + dz^2. \quad (13)$$

The convention being used here is that Roman indices run over spatial coordinates, while Greek indices run over time and space coordinates. To leading order in $\mathbf{a}$ and $\boldsymbol{\omega}$ the non-vanishing Christoffel symbols are:

$$\Gamma^i_{tt} \simeq -g_{tt,i} = a^i + (\boldsymbol{\omega}\times(\boldsymbol{\omega}\times\mathbf{x}))^i$$
$$\Gamma^i_{tj} \simeq \frac{1}{2}(g_{ti,j} - g_{tj,i}) = -c\,\epsilon_{ijk}\omega^k. \quad (14)$$

and the geodesic equations yield

$$\frac{d^2\mathbf{x}}{dt^2} = -\mathbf{a} - 2\boldsymbol{\omega}\times\mathbf{v} - \boldsymbol{\omega}\times(\boldsymbol{\omega}\times\mathbf{x}), \quad (15)$$

which recovers the form of the acceleration attributed to pseudo forces in a non-inertial frame. Turning this around, it is always possible to find a coordinate transformation to an inertial frame where the pseudo forces vanish. But this does not mean that we can simply transform gravity away.

2.3 *Tides and Curvature*

Einstein's trick for making gravity vanish only works in a small region of spacetime. It is impossible to remove the tidal forces that manifest over larger regions.

Suppose we do an experiment in a laboratory on the surface of the Earth as shown in Fig. 4. We can set up a coordinate system where the z axis points in the outward radial direction at the center of the lab, and the x, y directions span the floor of the lab. Now suppose that we release two masses from near the ceiling of the lab, the first with position vector relative to the center of the Earth given by $\mathbf{r}_1 = \mathbf{R} + \mathbf{x}_1$, where $\mathbf{R}$ is a vector connecting the center of the Earth to the floor of the lab, and the second with $\mathbf{r}_2 = \mathbf{R} + \mathbf{x}_2$. Initially the distance between the two masses is $L = |\Delta x| = |\mathbf{x}_2 - \mathbf{x}_1| = y_2 - y_1$. To leading order in $L/R_\oplus$ Newton's theory of gravity predicts a constant acceleration in the $-z$ direction:

$$\frac{d^2\mathbf{x}_{1,2}}{dt^2} = -\frac{GM_\oplus}{R_\oplus^2}\hat{z} = -g\hat{z}\,. \tag{16}$$

This part of the gravitational field can be transformed away by adopting a freely falling reference frame. Continuing the expansion of the Newtonian equations of motion to next order we encounter tidal forces in the y direction:

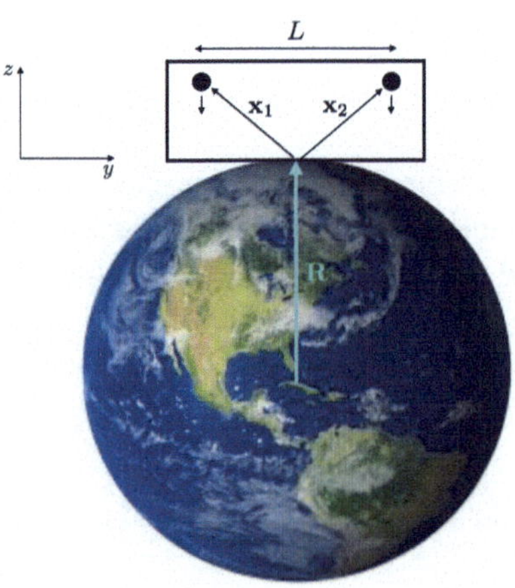

Fig. 4 Tidal forces in an Earth bound laboratory

$$\frac{d^2\mathbf{\Delta x}}{dt^2} = -\frac{GM_\oplus L}{R_\oplus^3}\hat{y}\,. \tag{17}$$

These non-uniform accelerations cannot be transformed away. Similarly, if one were to consider the motion of a single mass over an extended period of the time the local value of the acceleration g would change with time, and this change in acceleration can not be transformed away. In summary, the effects of gravity can only be transformed away across small regions of space for a short period of time.

Looking at the geodesic equation (11), we see that transforming away local acceleration terms is equivalent to setting the first derivatives of the metric equal to zero. It turns out that we always have enough coordinate freedom to set the components of the metric in the neighborhood of an event equal to the Minkowski metric $\eta_{\mu\nu} = \text{diag}(-1, 1, 1, 1)$, *and* to set the first derivatives equal to zero. However, there is not enough coordinate freedom to remove the second and higher derivatives. As shown by Riemann, the second derivatives of the metric describe the curvature of the spacetime. The components of the Riemann curvature tensor are given by

$$R^{\kappa}_{\mu\lambda\nu} = \Gamma^{\kappa}_{\mu\lambda,\nu} - \Gamma^{\kappa}_{\mu\nu,\lambda} + \Gamma^{\alpha}_{\mu\nu}\Gamma^{\kappa}_{\alpha\lambda} - \Gamma^{\alpha}_{\mu\lambda}\Gamma^{\kappa}_{\alpha\nu}\,. \tag{18}$$

The Christoffel symbols $\Gamma^{\alpha}_{\mu\nu}$, defined in Eq. (11), involve first derivatives of the metric and can be made to vanish at any point in spacetime, however their derivatives can not. In local free fall coordinates the metric components take the form

$$g_{\mu\nu} = \eta_{\mu\nu} - \frac{1}{3}R_{\mu\alpha\nu\beta}\Delta x^{\alpha}\Delta x^{\beta} + \cdots \tag{19}$$

This is nothing other than a Taylor series expansion about a point P, where the coordinates have been chosen such that $g_{\mu\nu,\lambda}|_P = 0$. These are variously refers to as free fall, locally Lorentzian or Riemann normal coordinates. These coordinates can be extended along the worldline of a particle to yield a locally non-rotating inertial frame defined by a set of four orthogonal basis vectors $\mathbf{e}_{(\gamma)}$. Here γ labels the basis vectors and should not be confused with the components of the basis vector which are labelled by a superscript: $e^{\alpha}_{(\gamma)}$. The locally non-rotating frame is carried along the worldline of the particle by Fermi–Walker transport:

$$u^{\beta}\nabla_{\beta}e^{\alpha}_{(k)} = e^{\alpha}_{(t)}g_{\mu\nu}e^{\mu}_{(k)}u^{\beta}\nabla_{\beta}e^{\nu}_{(t)}\,. \tag{20}$$

Fermi–walker transport keeps the basis vector $\mathbf{e}_{(t)}$ tangent to the worldline, as shown in Fig. 5.

While Fermi–Walker transport can eliminate "pseudo forces" along the worldline of a single particle, spacetime curvature prevents us from extending this inertial coordinate system globally. Spacetime curvature manifest as a tidal force that causes initially parallel geodesics to converge, diverge or twist about one another. Considering the geodesic equation for two nearby geodesics with separation vector $\boldsymbol{\xi}$. We find that spacetime curvature causes a relative acceleration:

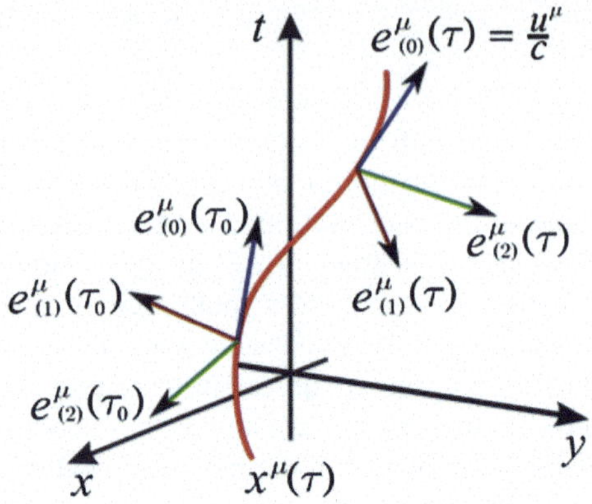

Fig. 5 Fermi–Walker transport of a locally inertial coordinate system along the worldline of a particle $x^\mu(\tau)$ parametrized by proper time τ

$$\frac{d^2\xi^\mu}{d\lambda^2} = u^\alpha\nabla_\alpha(u^\beta\nabla_\beta\xi^\mu) = R^\mu_{\ \alpha\beta\gamma}u^\alpha u^\beta\xi^\gamma\,. \tag{21}$$

In the slow motion limit where $u^t \gg u^i$ the equation for geodesic deviation becomes

$$\frac{d^2\xi^i}{dt^2} = c^2 R^i_{\ 0j0}\,\xi^j = R^i_{\ tjt}\,\xi^j\,. \tag{22}$$

Comparing with the expression (17), the tidal gravitational field of the Earth implies a spacetime curvature with $R^y_{\ tyt} = -GM_\oplus/R^3_\oplus$. Shortly we will see that this exactly what is predicted by Einstein's general theory of relativity.

2.4 Newtonian Gravity in Geometric Form

Before moving to the full Einstein equations, it is interesting to cast Newtonian gravity in geometrical terms. Consider a spacetime with line element

$$ds^2 = -(c^2 + 2\Phi(\mathbf{x}))dt^2 + dx^2 + dy^2 + dz^2\,, \tag{23}$$

where $\Phi \ll c^2$ is the Newtonian gravitational potential derived from Poisson's equation $\nabla^2\Phi = 4\pi G\rho$, where ρ is the mass density. The non-vanishing components of the Christoffel symbol are:

$$\Gamma^i_{tt} = \Phi_{,i} \qquad \Gamma^t_{ti} = \Gamma^t_{it} = \frac{\Phi_{,i}}{(c^2 + 2\Phi(\mathbf{x}))}\,. \tag{24}$$

The geodesic equation reduces to the usual Newtonian force law

$$\frac{d^2\mathbf{x}}{dt^2} = -\nabla\Phi\,, \tag{25}$$

and the non-vanishing components of the Riemann tensor are given by

$$R^i_{tjt} = -\Gamma^i_{tt,j} = \Phi_{,ij} \tag{26}$$

Returning to the Earth bound lab example, we have $\Phi = GM_\oplus/r$, $r^2 = R_\oplus^2 + y^2$ and

$$\Phi_{,yy} = \frac{2GM_\oplus}{r^3}(r_{,y})^2 - \frac{GM_\oplus}{r^2}(r_{,yy}) \approx -\frac{GM_\oplus}{R_\oplus^3}\,, \tag{27}$$

which recovers our earlier result. We see that in the Newtonian limit the Riemann tensor is nothing other than the tidal tensor of the Newtonian potential. The metric (23) accurately describes all aspects of Newtonian gravity, and corresponds to the the weak field *and* slow motion approximation to general relativity. It cannot, therefore, be used to model relativistic effects such as the deflection of starlight passing near the Sun (the prediction using (23) turns out to off by a factor of two).

2.5 Einstein Equations

The geometric treatment of Newtonian gravity has shown that curvature is related to the tidal field, and that in the weak field limit

$$R^j_{tjt} = g^{ij}\Phi_{,ij} = \nabla^2\Phi = 4\pi G\rho. \tag{28}$$

The quantity on the right-hand side of this equation is the tt component of the Ricci tensor $R_{\mu\nu} = R^\kappa_{\mu\kappa\nu}$, while the left hand side is proportional to the tt component of the stress-energy-momentum tensor $T_{\mu\nu}$. We seek to generalize (28) using terms that involve at most second derivatives of the metric. Moreover, we seek a coordinate invariant expression rather than one that singles out specific components of a tensor. The most general expression of this form is

$$\mathbf{R} + \alpha R\mathbf{g} + \Lambda\mathbf{g} = \kappa\mathbf{T} \tag{29}$$

where α, Λ and κ are constants, $R = \text{trace}(\mathbf{R})$ is the Ricci scalar and the bold-faced letters denote the Ricci, metric and energy-momentum tensors in coordinate-free notation. Conservation of energy-momentum $\nabla\cdot\mathbf{T} = 0$ requires that $\alpha = -1/2$, and recovery of the Newtonian limit (28) fixes $\kappa = 8\pi G/c^4$. The Einstein equations in component form are then [19]

$$R_{\mu\nu} - \frac{1}{2}g_{\mu\nu}R + \Lambda g_{\mu\nu} = \frac{8\pi G}{c^4}T_{\mu\nu}\,. \tag{30}$$

The quantity Λ is the cosmological constant. A remarkable feature of Einstein's theory is that the field equations can be used to derive the equations of motion of objects in spacetime [20, 21]. The derivation involves solving Einstein's equations in the limit of a small concentration of mass moving in the geometry generated by a larger concentration of mass [44]. Work continues on this problem today as part of the self-force program, with the goal of deriving waveforms describing small compact objects spiraling into massive black holes—otherwise known as Extreme Mass Ratio Inspires, or EMRIs [53]. The current state of the art yields equations of motion that are valid to first order in the mass ratio of the two bodies [25]:

$$u^\mu \nabla_\mu u^\nu = \frac{1}{2M} R_{\alpha\beta\gamma}{}^{\nu} S^{\alpha\beta} u^\gamma - (g^{\nu\kappa} + u^\nu u^\kappa) \left(\nabla_\alpha h^{\rm tail}_{\kappa\gamma} - \frac{1}{2} \nabla_\kappa h^{\rm tail}_{\gamma\alpha} \right) u^\gamma u^\alpha . \tag{31}$$

Here $S^{\alpha\beta}$ is the spin tensor for the small body, and $h^{\rm tail}_{\kappa\gamma}$ are the "tail terms" of the gravitational waves produced by the motion of the small body. The tail terms are proportional to the mass of the smaller body, and arise from the failure of Huygens principle in curved spacetime. The tail terms depend on the entire past history of the motion. To lowest order, ignoring the mass and spin of the small body, we see that small objects follow geodesics of the spacetime. The spin-curvature terms are named after Papapetrou and Dixon (though they were not the first to discover them). The self-force terms are called the MiSaTaQuWa equations, and were primarily derived by Mino et al. [39], Quinn and Wald [46].

2.6 Black Holes

The Einstein equations (30) applied to a fully general spacetime metric represent ten coupled, non-linear partial differential equations. Solving such equations is extremely challenging, even by numerical means. The equations become more tractable when applied to spacetimes with a high degree of symmetry. One of the earliest exact solutions to Einstein's equations was found by Schwarzschild for the case of a vacuum static, spherically symmetric spacetime with line element

$$ds^2 = -U(r)c^2 dt^2 + V(r)dr^2 + r^2 d\theta^2 + r^2 \sin^2\theta d\phi^2 . \tag{32}$$

The tt and rr components of the vacuum Einstein equations become

$$\frac{V'}{rV^2} + \frac{1}{r^2}\left(1 - \frac{1}{V}\right) = 0 \tag{33}$$

$$\frac{U'}{rU} - \frac{V}{r^2}\left(1 - \frac{1}{V}\right) = 0 \tag{34}$$

where $A' = dA/dr$. Equation (33) implies that $V^{-1} = (1 - R_s/r)$ where R_s is called the Schwarzschild radius. Combining rV times Eq. (33) with r times Eq. (34) yields $(\ln(UV))' = 0$, so that $U = V^{-1}$ (up to an arbitrary constant that be absorbed by rescaling t). To recover the Newtonian limit (23) for large r we have to set the Schwarzschild radius proportional to the mass of the central object: $R_s = 2GM/c^2$, yielding the Schwarzschild line element

$$ds^2 = -\left(1 - \frac{2GM}{c^2 r}\right) c^2 dt^2 + \frac{dr^2}{\left(1 - \frac{2GM}{c^2 r}\right)} + r^2 d\theta^2 + r^2 \sin^2\theta d\phi^2 \,. \tag{35}$$

For an object such as the Sun, the Schwarzschild radius $R_{\odot,s} = 2.95$ km would lie deep within the interior of the Sun, where the vacuum solution is no longer valid.

For a pure vacuum spacetime, the metric (35) is singular at the Schwarzschild radius $r = R_s$ and at $r = 0$. However, coordinate singularities are not physical, and the spacetime can look quite different in alternative coordinate systems. Note that general coordinate transformation have the form

$$x^{\bar{\mu}} = \Lambda^{\bar{\mu}}_{\nu} x^{\nu} = \frac{\partial x^{\bar{\mu}}}{\partial x^{\nu}} x^{\nu} \tag{36}$$

with tensor components transforming as

$$A^{\bar{\mu}}_{\bar{\nu}} = \Lambda^{\bar{\mu}}_{\alpha} \Lambda^{\beta}_{\bar{\nu}} A^{\alpha}_{\beta} \tag{37}$$

and similarly for higher rank tensors. The coordinate transformation $r = r'(1 + R_s/(4r'))^2$ yields the metric in isotropic form where the singularity at $r = R_s$ moves to a singularity at $r' = 4R_s$:

$$ds^2 = -\left(\frac{1 - \frac{R_s}{4r'}}{1 - \frac{R_s}{4r'}}\right)^2 c^2 dt^2 + \left(1 + \frac{R_s}{4r'}\right)^4 \left[dr'^2 + r'^2 d\theta^2 + r'^2 \sin^2\theta d\phi^2\right] . \tag{38}$$

Going a step further, the coordinate transformation $r = (3(R - cT)/2)^{2/3} R_s^{1/3}$ yields the Lemaître form of the metric which is only singular at $r = 0$:

$$ds^2 = -c^2 dT^2 + \frac{R_s}{r} dR^2 + r^2 d\theta^2 + r^2 \sin^2\theta d\phi^2 \,. \tag{39}$$

The nature of the surface at $r = R_s$ was not understood for many decades, and even today debate continues about what happens to quantum fields and superstrings in this spacetime leading to the suggestion of exotic phenomena such as "firewalls" or "fuzzballs" [38]. What is now understood at least for classical (non-quantum) objects is that geodesics can be continued through the Schwarzschild surface, and the tidal forces are finite there. One way to see this is to compute the components of the curvature tensor using an orthonormal coordinate system, denoted by hats, $\hat{\mu}$, where locally $g_{\hat{\mu}\hat{\nu}} = \eta_{\mu\nu}$. The non-vanish components of the Riemann tensor are then

$$R_{\hat{r}\hat{t}\hat{r}\hat{t}} = -R_{\hat{\theta}\hat{\phi}\hat{\theta}\hat{\phi}} = 2R_{\hat{\theta}\hat{t}\hat{\theta}\hat{t}} = 2R_{\hat{\phi}\hat{t}\hat{\phi}\hat{t}} = -2R_{\hat{r}\hat{\phi}\hat{r}\hat{\phi}} = -2R_{\hat{r}\hat{\theta}\hat{r}\hat{\theta}} = \frac{R_s}{r^3}\,. \tag{40}$$

We see that the curvature is finite at $r = R_s$ and divergent at $r = 0$. While non-singular, the Schwarzschild surface does have many special properties. For example, the redshift of a photon sent from a static source at radius r to a distant observer is given by

$$z(r) = \frac{1}{\sqrt{1 - \frac{R_s}{r}}} - 1\,. \tag{41}$$

We see that $r = R_s$ defines a surface of infinite redshift. Moreover, the force required to stay at a fix radius r also diverges at $r = R_s$:

$$F^{\hat{r}} = \frac{mR_s}{2\sqrt{1 - \frac{R_s}{r}}}\,. \tag{42}$$

Further insight can be gained by considering photon geodesics, which reveal the causal structure of the spacetime. In particular, radial null geodesics define the past and future light cones, as all other geodesics (null or timeline) lie within these cones. The condition $\mathbf{u} \cdot \mathbf{u} = 0$ for radial null geodesics yields the relation

$$\frac{dr}{dt} = \pm c\left(1 - \frac{R_s}{r}\right)\,. \tag{43}$$

Introducing the Regge–Wheeler radial coordinate $r_* = r + R_s \ln |r/R_s - 1|$ we find $dr_*/dt = \pm 1$, so the light cones are given by $t \pm r_* = \text{const}$. Adopting the new coordinate $v = ct + r_*$, which is constant along ingoing null geodesics, we arrive at the Eddington–Finkelstein form of the metric

$$ds^2 = -\left(1 - \frac{R_s}{r}\right)dv^2 + 2dvdr + r^2 d\theta^2 + r^2 \sin^2\theta d\phi^2\,. \tag{44}$$

Ingoing null geodesics have $ct - r_* = \text{const.}$, which in the Eddington–Finkelstein becomes $v = 2r + 2R_s \ln |r/R_s - 1| + \text{const}$. Introducing the new time coordinate $t_* = v - r$ so that ingoing null geodesics are straight lines at $-45°$ to the r axis, we can plot the inward and outward null geodesics as shown in Fig. 6. We see that the ingoing radial null geodesics cross the event horizon and terminate on the central singularity. Outgoing null geodesics that outside $r = R_s$ continue to travel outward in r, while those inside the Schwarzschild radius are trapped, and destined to encounter the singularity at $r = 0$. Since null geodesics define the light cones for all time-

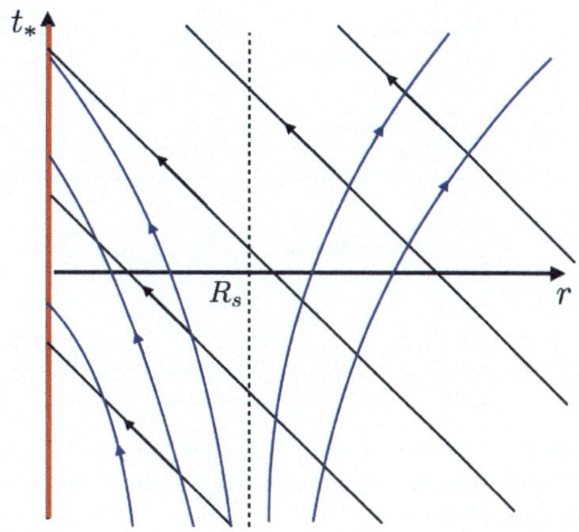

Fig. 6 Ingoing (black) and outgoing (blue) null geodesics in the Schwarzschild spacetime. All geodesics (null and otherwise) inside $r = R_s$ are unable to reach $r > R_s$, and are destined to encounter the curvature singularity at $r = 0$ (shown in red)

like and null paths—geodesics or otherwise—we see that $r = R_s$ defines a trapped surface—known as the event horizon—from which there is no escape.

Generic geodesics of the Schwarzschild spacetime can be derived by integrating the geodesics equations, but it is simpler to make use of the symmetries of the spacetime which imply the existence of several conserved quantities. The metric is invariant along the integral curves of the four Killing vectors $\boldsymbol{\xi}_1 = \boldsymbol{\partial}_t, \boldsymbol{\xi}_2 = \boldsymbol{\partial}_\phi, \boldsymbol{\xi}_3 = \sin\phi\boldsymbol{\partial}_\theta + \cot\theta\cos\phi\boldsymbol{\partial}_\phi$ and $\boldsymbol{\xi}_4 = \cos\phi\boldsymbol{\partial}_\theta - \cot\theta\sin\phi\boldsymbol{\partial}_\phi$. The symmetries tell us that $E = -\boldsymbol{\xi}_1 \cdot \mathbf{p}$, $L_z = \boldsymbol{\xi}_2 \cdot \mathbf{p}$, $L_3 = \boldsymbol{\xi}_3 \cdot \mathbf{p}$ and $L_4 = \boldsymbol{\xi}_4 \cdot \mathbf{p}$ are conserved quantities, where $\mathbf{p}$ is the four momentum. The four momentum also satisfies the normalization condition $\mathbf{p} \cdot \mathbf{p} = -m^2c^4$ for particles and $\mathbf{p} \cdot \mathbf{p} = 0$ for photons. We can use the rotational symmetry to place all orbits in the equatorial plane $\theta = \pi/2$, with $L_3 = L_4 = 0$. The normalization condition can then be expressed as

$$\frac{\tilde{E}^2 - 1}{2} = \frac{1}{2}\left(\frac{dr}{d\tau}\right)^2 + \tilde{V}(r) \quad \text{(particles)}$$
$$E^2 = \left(\frac{dr}{d\lambda}\right)^2 + V(r) \quad \text{(photons)} \tag{45}$$

where

$$\tilde{V}(r) = -\frac{M}{r} + \frac{\tilde{L}_z^2}{2r^2} - \frac{M\tilde{L}_z^2}{r^3},$$
$$V(r) = \frac{L^2}{r^2}\left(1 - \frac{2M}{r}\right). \tag{46}$$

play the role of effective potentials. Here $\tilde{E} = E/m$ and $\tilde{L}_z/m$ are the energy and angular momentum per unit mass. The expression for the radial velocity of a particle is identical to its Newtonian counterpart aside from the term $-M\tilde{L}_z^2/r^3$. This term has

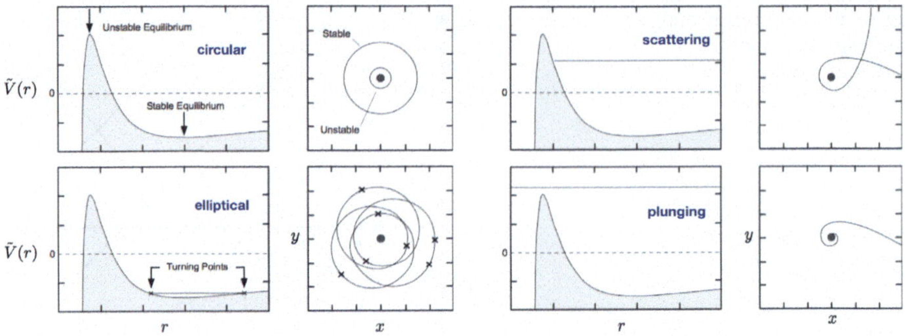

Fig. 7 Examples of particle geodesics in the Schwarzschild spacetime. The radial velocity can be read off from the difference in the height of the effective potential (shaded in grey) and the energy (horizontal lines)

important consequences for tightly bound orbits, allowing trajectories with $\tilde{L}_z \neq 0$ to reach $r = 0$, and giving rise to a collection of unstable circular orbits at $r_u = (\tilde{L}_z^2/R_s)(1 - \sqrt{1 - 3R_s^2/L_z^2})$. There is an inner most (marginally) stable circular orbit (ISCO) at $r = 3R_s$ that plays an important role in understanding accretion discs around black holes and the gravitational waveforms of merging black holes. Photon orbits can be described in an analogous way, though there are no bound orbits other than an isolated unstable circular orbit at $r = 3R_s/2$ (Fig. 7).

The Schwarzschild solution was generalized by Kerr to describe the asymptotically flat, stationary, axisymmetric spacetime that we interpret as a rotating black hole. The derivation is much more difficult, and the resulting spacetime has a much richer phenomenology. The metric in Boyer–Lindquist coordinates has the line element

$$ds^2 = \frac{-\Delta}{\Sigma}\left(cdt - a\sin^2\theta d\phi\right)^2 + \frac{\sin^2\theta}{\Sigma}\left((r^2 + a^2)d\phi - a\,cdt\right)^2 + \frac{\Sigma}{\Delta}dr^2 + \Sigma d\theta^2 \tag{47}$$

where

$$\begin{aligned} \Sigma &= r^2 + a^2\cos^2\theta, \\ \Delta &= r^2 - R_s r + a^2, \end{aligned} \tag{48}$$

and $a = S/(cM)$, where S is the spin angular momentum of the black hole and M the mass. The metric has coordinate singularities at $r_\pm = R_s/2 \pm \sqrt{R_s^2/4 - a^2}$. The surface at $r = r+$ is identified as the event horizon as no trajectories from inside r_+ can cross the surface. Additionally there is a static-limit surface with $r_{\rm SL} = R_s/2 + \sqrt{R_s^2/4 - a^2\cos^2\theta}$ interior to which it is impossible to stay fixed with respect to the distant stars—everything gets swept around in the swirling vortex of the black hole. Energy can be extracted from the black hole by scattering radiation or particles within the so-called ergo-region between r_+ and $r_{\rm SL}$. The spacetime has a curvature

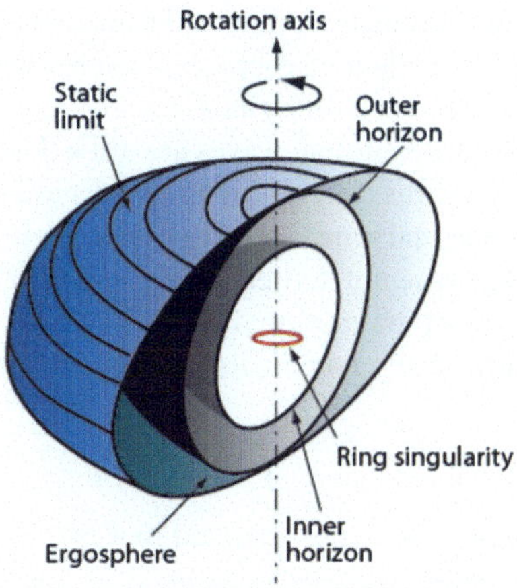

Fig. 8 Cross section showing the key features of a Kerr black hole

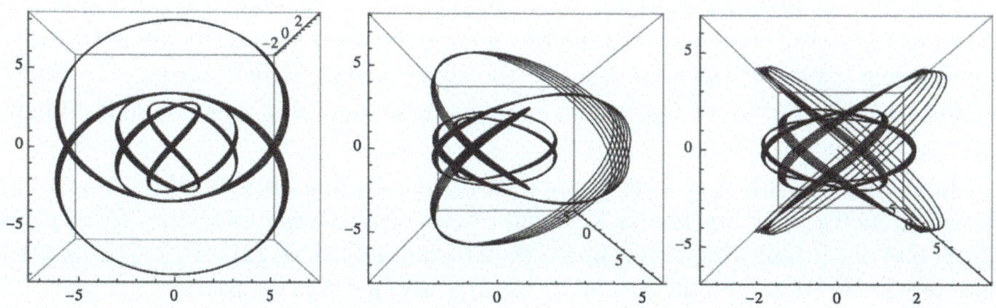

Fig. 9 Examples of the some of the highly non-Keplerian closed periodic orbits of the Kerr geometry

singularity along the ring $r = 0$, $\theta = \pi/2$. The key feature of a rotating black hole are illustrated in Fig. 8.

Geodesics in the Kerr geometry are fully specified by four constants of the motion that arise from the symmetries of the spacetime. The constant are the energy $E = -p_t$, the azimuthal angular momentum $L_z = p_\phi$, the mass $m^2 = -\mathbf{p} \cdot \mathbf{p}/c^4$ and the Carter constant $C = p_\theta^2 + p_\phi / \sin^2\theta$. The existence of the latter was something of a surprise, and its existence guarantees that the orbits are regular (non-chaotic). Closed orbits in Kerr can be highly non-Keplerian, exhibiting rich structure due to the presence of three distinct orbital frequencies associated with the radial, azimuthal and equatorial motion [36]. Examples of some closed periodic orbits of Kerr are shown in Fig. 9.

3 Gravitational Wave Theory

Gravitational waves are generated by flows of energy-momentum. When you wave to someone you are generating gravitational waves, though with amplitudes that are far to weak to be detected using existing technologies. The waves we can detect come from violent astrophysical events where large concentrations of mass move at close to the speed on light.

Gravitational waves are often described as "propagating ripples of curvature", and while this description is apt, deciding which part of the spacetime curvature to call a wave and what part to associate with the background spacetime is a subtle question that was not settled until the late 1960s [29, 30]. Here I will consider the simpler case of gravitational waves in a background Minkowski spacetime, where the waves are wholly responsible for the curvature. The metric and its inverse take the form

$$g_{\mu\nu} = \eta_{\mu\nu} + h_{\mu\nu}, \qquad g^{\mu\nu} = \eta^{\mu\nu} - h^{\mu\nu}, \tag{49}$$

where we have assumed that $|h_{\mu\nu}| \ll 1$, so that $h^{\mu\nu} = \eta^{\mu\kappa}\eta^{\nu\lambda}h_{\kappa\lambda}$. Note that it only makes sense to say that $|h_{\mu\nu}| \ll 1$ in a coordinate system where $\boldsymbol{\eta} = \text{diag}(-1, 1, 1, 1)$. For example, we may use cartesian coordinates or more generally, any orthonormal coordinate system. The linearized metric (49) can be used to describe more that just gravitational waves. It can also describe the solar system, which has $|h_{\mu\nu}| < 10^{-5}$, and the Universe out to ~1 Gpc (in an average sense, avoiding regions around black holes).

There are two classes of coordinate transformation that preserve the form of the linearized metric. The first are background Lorentz transformations $x^{\bar{\mu}} = L^{\bar{\mu}}_{\nu} x^{\nu}$, and the second are infinitesimal coordinate transformations (also called gauge transformations) $x^{\bar{\mu}} = x^{\mu} + \zeta^{\mu}$ with $|\zeta^{\mu}| \ll 1$. Using (36) and (37) we find

$$h_{\bar{\mu}\bar{\nu}} = h_{\mu\nu} - \partial_{\mu}\zeta_{\nu} - \partial_{\nu}\zeta_{\mu} \,. \tag{50}$$

It is a straightforward exercise to verify that the Riemann tensor, which has components

$$R_{\alpha\beta\mu\nu} = \frac{1}{2}(\partial_{\beta}\partial_{\mu}h_{\alpha\nu} + \partial_{\alpha}\partial_{\nu}h_{\beta\mu} - \partial_{\beta}\partial_{\nu}h_{\alpha\mu} - \partial_{\alpha}\partial_{\mu}h_{\beta\nu}) \,, \tag{51}$$

is invariant under these transformations. The linearized Einstein equations take the form

$$-\Box\bar{h}_{\mu\nu} + \partial_{\nu}\partial^{\alpha}\bar{h}_{\mu\alpha} + \partial_{\mu}\partial^{\alpha}\bar{h}_{\nu\alpha} - \eta_{\mu\nu}\partial^{\alpha}\partial^{\beta}\bar{h}_{\alpha\beta} = \frac{16\pi G}{c^4}T_{\mu\nu} \tag{52}$$

where we have introduced the trace-reversed metric perturbation

$$\bar{h}_{\mu\nu} = h_{\mu\nu} - \frac{1}{2}h\,\eta_{\mu\nu} \,, \tag{53}$$

(so-called because $\bar{h} = -h$), and the wave operator

$$\Box = \partial^{\alpha}\partial_{\alpha} = -\frac{1}{c^2}\frac{\partial^2}{\partial t^2} + \nabla^2 \,. \tag{54}$$

The linearized Einstein equations can be brought into a simpler form by utilizing some of the gauge freedom we have at our disposal. Under a gauge transformation the divergence of the trace-reverse metric transforms as

$$\partial^{\bar{\alpha}} \bar{h}_{\bar{\mu}\bar{\alpha}} = \partial^{\alpha} \bar{h}_{\mu\alpha} - \Box \zeta_{\mu} \,. \tag{55}$$

We can choose ζ_μ such that the right hand side of the above equation vanishes, resulting in the Lorentz family of gauges with $\partial^{\bar{\alpha}} \bar{h}_{\bar{\mu}\bar{\alpha}} = 0$. Note that some gauge freedom remains as we can add to ζ_μ any homogeneous solution of the wave equation $\Box \lambda_\mu = 0$ and still maintain the divergence free condition. Dropping primes on the coordinate indices, the Einstein equations in the Lorentz gauge become

$$\Box \bar{h}_{\mu\nu} = -\frac{16\pi G}{c^4} T_{\mu\nu} \,, \qquad \partial^{\alpha} \bar{h}_{\mu\alpha} = 0 \,. \tag{56}$$

These equations are very similar to the Maxwell equations in the Lorentz gauge,

$$\Box A^{\mu} = \mu_0 J^{\mu} \,, \qquad \partial_{\alpha} A^{\alpha} = 0 \,, \tag{57}$$

and be solved using similar techniques (Green's functions, expansion in orthogonal function—a.k.a. the full Jackson [32]).

3.1 *Newtonian Limit Redux*

Before continuing with the discussion of gravitational waves, let us pause for a moment and consider the weak-field, slow motion limit of the linearized Einstein equations, which appears as the first step in the post-Newtonian expansion of Einstein's equations. In the limit where $\partial_t^2 \ll c^2 \nabla^2$, $|T_{tt}| \gg |T_{ti}| \gg |T_{ij}|$ and objects are moving at much less than the speed of light, the linearized field equations reduce to

$$\Box \bar{h}_{tt} = -\frac{16\pi G}{c^4} T_{tt} \quad \Rightarrow \quad \nabla^2 \bar{h}_{tt} = -16\pi G \rho \,. \tag{58}$$

Making the identification $\bar{h}_{tt} = -4\Phi$ we have $\bar{h} = -h = 4\Phi$ and

$$ds^2 = -(c^2 + 2\Phi) dt^2 + \left(1 - \frac{2\Phi}{c^2}\right)(dx^2 + dy^2 + dz^2) \,. \tag{59}$$

This form for the metric is similar to what we encountered earlier in the geometrical description of Newtonian gravity (23), but includes an additional post-Newtonian term in the spatial metric. The normalization of 4-velocity $\mathbf{u} \cdot \mathbf{u} = -c^2$ implies that

$$u^t = \frac{dt}{d\tau} = 1 + \frac{v^2}{2c^2} - \frac{\Phi}{c^2} \,, \tag{60}$$

which includes the effects of relativistic time dilation and the slowing of clocks in a gravitational field.

3.2 Waves in Vacuum

Consider a plane wave expansion of the vacuum ($T_{\mu\nu} = 0$) field equations by writing $\bar{h}_{\mu\nu} = \Re\{A_{\mu\nu} e^{i\mathbf{k}\cdot\mathbf{x}}\}$, where $A_{\mu\nu}$ is a constant polarization tensor and $\mathbf{k} \rightarrow (\omega/c, k^i)$ is the wave vector. The linearized field Eqs. (57) yield the conditions

$$k^\alpha k_\alpha = -\omega^2 + k_i k^i = 0\,, \qquad A_{\mu\nu} k^\nu = 0\,. \tag{61}$$

The first condition tells us that gravitational waves travel at the speed of light, while second condition tells that in our chosen gauge, the oscillations are transverse to the direction of propagation. The polarization tensor is symmetric, so has ten components in a four dimensional spacetime. The transverse condition provides four constraints, so that six degrees of freedom remain. We can use the residual coordinate freedom $\zeta^\mu \rightarrow \zeta^\mu + \lambda^\mu$, where λ^μ is a homogenous solution to the wave equation, $\Box \lambda^\mu = 0$, to eliminate a further four degrees of freedom. Writing $\lambda^\mu = i C^\mu e^{i\mathbf{k}\cdot\mathbf{x}}$ we have

$${}^{(\mathrm{new})}A_{\mu\nu} = {}^{(\mathrm{old})}A_{\mu\nu} + C_\mu k_\nu + C_\nu k_\mu - \eta_{\mu\nu} k^\alpha C_\alpha\,. \tag{62}$$

We can use this gauge freedom in many different ways. One popular choice is the *Transverse-Traceless* (TT) gauge, where C_μ is chosen so as to make the polarization trace-free and orthogonal to the worldlines of timelike particles:

$${}^{(\mathrm{new})}A^\mu_\mu = 0 \qquad {}^{(\mathrm{new})}A_{\mu\nu} u^\mu = 0\,. \tag{63}$$

The trace-free condition sets one constraint, while the orthogonality condition sets three constraints, not four, since we already have that ${}^{(\mathrm{new})}A_{\mu\nu} u^\mu k^\nu = 0$ from the waves being transverse.

Consider the case of a plane gravitational wave propagating in the $+z$ direction, $\mathbf{k} \rightarrow (\omega/c, 0, 0, \omega)$, as seen by a stationary observer, $\mathbf{u} \rightarrow (c, 0, 0, 0)$. The TT gauge conditions (63) imply that $A_{t\mu} = A_{z\mu} = 0$ and $A_{yy} = -A_{xx}$. Writing $h_{xx} = h_+$ and $h_{xy} = h_\times$ we have

$$ds^2 = -c^2 dt^2 + (1 + h_+) dx^2 + (1 - h_+) dy^2 + 2h_\times dx dy + dz^2 \tag{64}$$

with

$$h_+ = A_+ \cos(\omega(t - z/c) + \phi_+), \qquad h_\times = A_\times \cos(\omega(t - z/c) + \phi_\times). \tag{65}$$

The plane wave spacetime (64) has a number of interesting properties, the most surprising being that objects at rest remain at rest in this coordinate system. This follows immediately from the geodesic equation

$$\frac{du^\alpha}{d\tau} = -\Gamma^\alpha_{\mu\nu} u^\mu u^\nu = \frac{1}{2}\partial^\alpha h_{\mu\nu} u^\mu u^\mu \,. \tag{66}$$

An initially stationary object has $u^\alpha = \delta^\alpha_t$, and since $\partial^\alpha h_{tt} = 0$, we see that the coordinate acceleration vanishes: $du^\alpha/d\tau = 0$. While objects stay at the same spatial coordinate locations, it does not mean that the waves have no measurable effect. For example, the curvature tensor has the non-vanishing components

$$\begin{aligned} R_{ytyt} = R_{yzyz} = R_{xtxz} = -R_{xtxt} = -R_{xzxz} = -R_{ytyz} &= \frac{1}{2}\ddot{h}_+ \\ R_{xtyz} = R_{ytxz} = -R_{xzyz} = -R_{xtyt} = -R_{ytxt} = -R_{yzxz} &= \frac{1}{2}\ddot{h}_\times \,, \end{aligned} \tag{67}$$

which tells us that the proper separation between nearby particles will vary with time, even though their coordinate locations remain fixed in space. What has happened here is that the wave has been put into the coordinates when we added in a solution of the homogeneous wave equation.

The TT gauge is by far the simplest gauge to work in when computing things like the response of a gravitational wave detector, but it does have the unfortunate property that the wave motion has been hidden inside the coordinates. The physical properties are more readily seen by transforming to a locally inertial frame. Recall that in Fermi Normal (FN) coordinates the metric can be written as $g_{\mu\nu} = \eta_{\mu\nu} - \frac{1}{3}R_{\mu\alpha\nu\beta}x^\alpha x^\beta$, which for the case at hand yields

$$\begin{aligned} ds^2 &\approx -c^2 d\bar{t}^2(1 + R_{titj}\bar{x}^i\bar{x}^j) - \frac{4}{3}d\bar{t}d\bar{x}^i(R_{tjik}\bar{x}^j\bar{x}^k) + d\bar{x}^i d\bar{x}^j\left(\delta_{ij} - \frac{1}{3}R_{ikjl}\bar{x}^k\bar{x}^l\right) \\ &= -c^2 d\bar{t}^2 + d\bar{x}^2 + d\bar{y}^2 + d\bar{z}^2 + \left(\ddot{h}_\times\bar{x}\bar{y} + \frac{1}{2}\ddot{h}_+(\bar{x}^2 - \bar{y}^2)\right)(c d\bar{t} - d\bar{z})^2 . \end{aligned} \tag{68}$$

In deriving this expression we have used the fact that the components of the Riemann tensor are unchanged to leading order in h by the coordinate transformation that takes us to the FN coordinate system. The coordinate transformation between TT and FN coordinates is, to leading order in h, given by Rakhmanov [47]

$$\begin{aligned} x &= \bar{x} - \frac{1}{2}h_+\bar{x} - \frac{1}{2}h_\times\bar{y} - \frac{1}{2}\bar{z}(\bar{x}\dot{h}_+ + \bar{y}\dot{h}_\times) \\ y &= \bar{y} + \frac{1}{2}h_+\bar{y} - \frac{1}{2}h_\times\bar{x} + \frac{1}{2}\bar{z}(\bar{y}\dot{h}_+ - \bar{x}\dot{h}_\times) \\ z &= \bar{z} + \frac{1}{4}(\bar{x}^2 - \bar{y}^2)\dot{h}_+ + \frac{1}{2}\bar{x}\bar{y}\dot{h}_\times \\ t &= \bar{t} - \frac{1}{4}(\bar{x}^2 - \bar{y}^2)\dot{h}_+ - \frac{1}{2}\bar{x}\bar{y}\dot{h}_\times \,. \end{aligned} \tag{69}$$

With a little algebra it is easy to show that the line element (64) is transformed to the line element (68) under the coordinate transformation (69). Remarkably, while the construction of the FN coordinate system is usually only valid locally, the metric (69) turns out to be an exact solution of Einstein's equations that is valid globally [47], which is useful when considering detectors where the arm lengths are large compared to the wavelength of the gravitational wave (as is the case for pulsar timing).

Restricting our attention for now to the long wavelength limit, were the wavelength of the gravitational wave is much greater than the size of the detector: $|\bar{x}|, |\bar{y}| \ll \lambda = 2\pi c/\omega$, the geodesic equation in the FN metric yields

$$\begin{aligned}\frac{d^2\bar{x}}{d\bar{t}^2} &= \frac{1}{2}\bar{x}\ddot{h}_+ + \frac{1}{2}\bar{y}\ddot{h}_\times \\ \frac{d^2\bar{y}}{d\bar{t}^2} &= \frac{1}{2}\bar{x}\ddot{h}_\times - \frac{1}{2}\bar{y}\ddot{h}_+ \\ \frac{d^2\bar{z}}{d\bar{t}^2} &= 0\,. \end{aligned} \tag{70}$$

We see that particles oscillate back and forth in the plane orthogonal to the propagation direction. A ring of test particles that is initially at rest in the plane $\bar{z} = 0$ with coordinates $\bar{x}(0) = L\cos\phi$, $\bar{y}(0) = L\sin\phi$ will oscillate as

$$\begin{aligned}\bar{x} &= L\left(\cos\phi + \frac{1}{2}(h_+\cos\phi + h_\times\sin\phi)\right) \\ \bar{y} &= L\left(\sin\phi + \frac{1}{2}(h_+\cos\phi - h_\times\sin\phi)\right)\end{aligned} \tag{71}$$

The motion is illustrated in Fig. 10.

3.3 *Making Waves*

The linearized Einstein equations (56) can be formally solved using the Green's function

$$G(\mathbf{x}-\mathbf{x}') = \frac{1}{4\pi|\mathbf{x}_s - \mathbf{x}'_s|}\delta(t_{\rm ret} - t'_{\rm ret})\,, \tag{72}$$

where $\mathbf{x}_s$ denotes the spatial part of the 4-vector $\mathbf{x}$ and $t_{\rm ret} = t - |\mathbf{x}_s - \mathbf{x}'_s|/c$ is the retarded time. The Green's function satisfies the equation $\Box G(\mathbf{x}-\mathbf{x}') = \delta^4(\mathbf{x}-\mathbf{x}')$. The formal solution is then

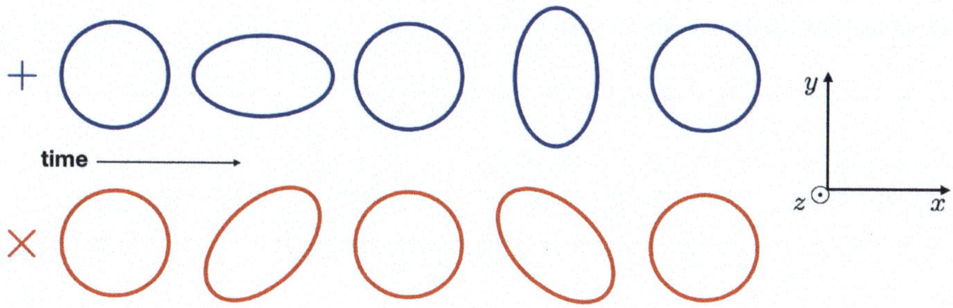

Fig. 10 The distortion of a ring of test particles caused by the plus and cross polarization states of a plane gravitational wave propagating in the z direction

$$\begin{aligned}\bar{h}_{\mu\nu}(\mathbf{x}) &= \frac{-16\pi G}{c^4}\int d^4x'\, G(\mathbf{x}-\mathbf{x}')T_{\mu\nu}(\mathbf{x}') \\ &= \frac{4G}{c^4}\int d^3x'\,\frac{T_{\mu\nu}(t_{\rm ret},\mathbf{x}'_s)}{|\mathbf{x}_s-\mathbf{x}'_s|}\,. \end{aligned} \tag{73}$$

The general solution can be expressed in the TT gauge by applying the projection tensor P_{ijkl}, which removes any longitudinal components and subtracts the trace: $h_{ij}^{\rm TT} = P_{ijkl}\bar{h}^{kl}$. The projection tensor is defined:

$$P_{ijkl} = p_{ik}p_{jl} - \frac{1}{2}p_{ij}p_{kl}\,, \quad \text{where} \quad p_{ij} = \delta_{ij} - n_i n_j\,, \tag{74}$$

and $\hat{n}$ is the unit vector in the direction of the source.

The general solution (73) is not particularly illuminating. To arrive at an expression that can be used in practice we need to make some approximations. The first is that the size of the source region d is very much smaller than the distance to field point r where we are evaluating the wave:

$$|\mathbf{x}_s - \mathbf{x}'_s| \approx r - \mathbf{x}'_s\cdot\hat{n} + \mathcal{O}\left(\frac{d^2}{r}\right), \tag{75}$$

so that

$$h_{ij}^{\rm TT} = \frac{4G}{rc^4}P_{ij}{}^{kl}\int d^3x'\,T_{kl}(t - r/c + \mathbf{x}'_s\cdot\hat{n}/c, \mathbf{x}'_s)\,. \tag{76}$$

The second approximation we make is that the material is moving slowly compared to the speed of light, which allows us to Taylor expand the energy-momentum tensor:

$$\begin{aligned} T_{kl}(t - r/c + \mathbf{x}'_s\cdot\hat{n}/c, \mathbf{x}'_s) = T_{kl}(t - r/c) &+ \frac{x'_i n^i}{c}\partial_t T_{kl}(t - r/c) \\ &+ \frac{x'_i x'_j n^i n^j}{2c^2}\,\partial_t^2 T_{kl}(t - r/c) + \cdots \end{aligned} \tag{77}$$

Define the multipole decomposition of the source:

$$
\begin{aligned}
S^{ij}(t) &= \int d^3x'\, T^{ij}(t, \mathbf{x}'_s) \\
S^{ijk}(t) &= \frac{1}{c} \int d^3x'\, T^{ij}(t, \mathbf{x}'_s) x'^k \\
S^{ijkl}(t) &= \frac{1}{c^2} \int d^3x'\, T^{ij}(t, \mathbf{x}'_s) x'^k x'^l
\end{aligned}
\tag{78}
$$

we can write

$$
\begin{aligned}
h_{ij}^{\mathrm{TT}} = \frac{4G}{rc^4} P_{ijkl} \big[& S^{kl}(t - r/c) + n_m \dot{S}^{klm}(t - r/c) \\
& + \frac{1}{2} n_m n_p \ddot{S}^{klmp}(t - r/c) + \cdots \big] .
\end{aligned}
\tag{79}
$$

For our purposes it will be enough to consider the lowest order in the multipole expansion, which turns out to be proportional to the second time derivative of the quadrupole moment: $S^{ij}(t) = \ddot{Q}^{ij}(t)/(2c^2)$. To establish this result we need to take time derivatives of the quadrupole moment and integrate by parts. For slow moving sources we have

$$
Q^{ij} = \int d^3x\, T_{tt}(t, \mathbf{x}_s) x^i x^j \approx \int d^3x \rho(t, \mathbf{x}_s) x^i x^j . \tag{80}
$$

Thus

$$
\begin{aligned}
\dot{Q}^{ij} &= \int d^3x\, \partial_t T_{tt} x^i x^j = c \int d^3x \partial^k T_{tk} x^i x^j \\
&= -c \int d^3x (T_t^i x^j + T_t^j x^i) + c \oint T^{tk} x^i x^j\, d^2S .
\end{aligned}
\tag{81}
$$

In the first line we have used conservation of energy to swap the time derivative for a spatial derivative. The surface integral on the second line vanishes since the surface is outside the source. Taking a second time derivative we have

$$
\begin{aligned}
\ddot{Q}^{ij} &= -\int d^3x\, c^2 (\partial_k T^{ki} x^j + \partial_k T^{kj} x^i) \\
&= 2c^2 \int d^3x T^{ij} - c^2 \oint (T^{ki} x^j + T^{kj} x^i) n_k d^2S .
\end{aligned}
\tag{82}
$$

Once again the surface integral vanishes, and we arrive at the promised result. Putting everything together we arrive at the leading order, quadrupole formula for gravitational wave emission:

$$
h_{ij}^{\mathrm{TT}} = \frac{2G}{rc^6} P_{ijkl} \ddot{Q}^{kl}(t - r) . \tag{83}
$$

Applying this expression to a gravitational wave traveling in the z direction we find

$$\begin{aligned} h_+ &= \frac{G}{rc^6}\left(\ddot{Q}_{xx}(t-r) - \ddot{Q}_{yy}(t-r)\right) \\ h_\times &= \frac{2G}{rc^6}\ddot{Q}_{xy}(t-r)\,. \end{aligned} \tag{84}$$

3.4 Energy and Momentum of a Gravitational Wave

Gravitational waves carry energy and momentum away from a source, and through the non-linearity of Einstein's equations, become sources that modify the background geometry and even generate waves of their own. The energy and angular momentum carried by gravitational waves causes binary stars to spiral inward and eventually merge. The linear momentum carried by gravitational waves can lead to recoil kicks during black hole mergers that send the merged black hole racing away at thousands of kilometers per hour. The calculation of the energy and momentum carried by gravitational waves raises several subtle issues that deserve a more careful treatment that can be squeezed into these lectures, so here I sketch out the main results.

The derivation begins by expanding the metric to next order:

$$g_{\mu\nu} = \eta_{\mu\nu} + h_{\mu\nu} + f_{\mu\nu} \tag{85}$$

where $|f_{\mu\nu}| \sim |h_{\mu\nu}|^2$. The Einstein equations are then expanded order-by-order:

$$G^{(1)}_{\mu\nu}(h) = \frac{8\pi G}{c^4}T_{\mu\nu} \qquad G^{(2)}_{\mu\nu}(f) = \frac{8\pi G}{c^4}\tau_{\mu\nu}(h^2)\,. \tag{86}$$

The leading order equation is what we considered earlier. The second order equation is sourced by the energy momentum tensor for gravitational waves, which in the TT gauge has the form

$$\tau^{\rm TT}_{\mu\nu} = \frac{c^4}{32\pi G}\langle \partial_\mu h^{\rm TT}_{jk}\partial_\nu h^{jk}_{\rm TT}\rangle\,. \tag{87}$$

The angle brackets denote an average over a region of spacetime that covers several wavelengths and wave cycles. The averaging is needed since gravitational energy cannot be localized. The gravitational wave energy momentum tensor is traceless $\tau^\mu_\mu = 0$ and conserved $\partial_\nu \tau^{\mu\nu} = 0$.

In spherical coordinates, and working to quadrupole order we have

$$\tau^{\rm TT}_{tt} = \tau^{\rm TT}_{rr} = -\tau^{\rm TT}_{tr} = \frac{c^4}{32\pi G}\langle|\dot{h}^{\rm TT}_{ij}|^2\rangle = \frac{c^2}{8\pi r^2 G}\langle|\dddot{Q}^{\rm TT}_{ij}|^2\rangle\,. \tag{88}$$

The energy radiated by a source is given by

$$\frac{dE}{dt} = \oint \tau_{tr}^{\rm TT} r^2 \sin^2\theta \, d\theta d\phi = \frac{1}{5} \langle | \dddot{Q}_{ij}^{\rm TT}(t-r)|^2 \rangle \,. \tag{89}$$

The linear momentum radiated and the angular momentum radiated can be calculated in a similar fashion:

$$\frac{dP^k}{dt} = -\frac{r^2}{32\pi} \oint d\Omega \, \langle \dot{h}_{ij}^{\rm TT} \partial^k h_{\rm TT}^{ij} \rangle \tag{90}$$

and

$$\frac{dJ^i}{dt} = \frac{r^2}{32\pi} \oint d\Omega \, \langle 2\epsilon^{ikl} \dot{h}_{al}^{\rm TT} \dot{h}_{ak}^{\rm TT} - \epsilon^{ikl} \partial^k h_{ab}^{\rm TT} x_k \partial_l h_{\rm TT}^{ab} \rangle \,. \tag{91}$$

4 Gravitational Wave Detection

There are three main techniques used to detect gravitational waves. Acoustic detectors, time-of-flight detectors and astrometry. Acoustic, or bar detectors, seek to measure the tidal force imparted by a gravitational wave through the alternate stretching and compression of a mechanical oscillator, such as a large bar of aluminum. Acoustic detectors are sensitive in a narrow frequency band around the resonant frequency of the oscillator, and this limitation, along with practical challenges in achieving high sensitivity, have seen bar detectors abandoned in favor of wide-band laser interferometers. Time-of-flight detectors come in many forms, and include ground and space interferometers, spacecraft doppler tracking and pulsar timing. While the measurement techniques differ, the underling observable is the same: the small changes in the time of arrival caused by gravitational waves. Astrometric detection is a relatively new approach that seeks to measure the apparent change in the arrival direction of starlight through gravitational lensing by gravitational waves. In these lectures I will only consider time-of-flight detectors as they are currently the most sensitive and widely used.

4.1 Photon Timing

The time that it takes a photon to propagate between two points in space will be perturbed by the presence of gravitational waves. Figure 11 illustrates the measurement principle behind pulsar timing, spacecraft doppler tracking and laser interferometers. The pulsar timing approach to gravitational wave detection operates directly on this principle. The highly regular radio pulses from a milli-second pulsar will arrive a little earlier or a little later than they would if no gravitational waves were perturbing the spacetime geometry. Spacecraft doppler tracking measures changes in the frequency of radio signals sent from Earth and transponded back from a satellite. Here the measurement is proportional to the time derivative of the photon propagation time. Laser interferometers measure the phase shifts imparted on a laser signal that is sent down two paths and reflected or transponded back to a common point where the phase of the two beams can be compared. The phase shift is directly proportional to the difference in propagation time long the two paths. To calculate the response of each detector type we only need to calculate the general expression for the change in propagation time caused by gravitational waves for photons propagating between two points in space. The response is then found by combining the effects along the entire photon path, which amounts to a single pass for pulsar timing, two passes for spacecraft doppler tracking and four passes for a Michelson interferometer. More complicated measurement paths, such as a Michelson interferometer with Fabry–Perot cavities, or a space interferometer using time delay interferometry, just require additional single passes to be added together.

In most applications the distance to the gravitational wave source is much larger than the distance along the arms of the detector, which allows us to describe the perturbed spacetime using the plane wave metric (64) that we considered earlier. Making the coordinate change $u = ct - z$, $v = ct + z$ the line element takes the form

$$ds^2 = -du dv + (1 + h_+(u))dx^2 + (1 - h_+(u))dy^2 + 2h_\times(u)dxdy + dz^2 \,. \quad (92)$$

Since we are working in the TT-gauge, particles at rest will remain at the same spatial coordinate location. Our goal is to compute the time it takes a photon to propagate from the spatial origin (0, 0, 0) to the point (x, y, z). The high degree of symmetry of the plane wave spacetime (92) makes computing geodesics simple [The derivation here follows [12, 22]]. The spacetime is invariant along the Killing vector fields

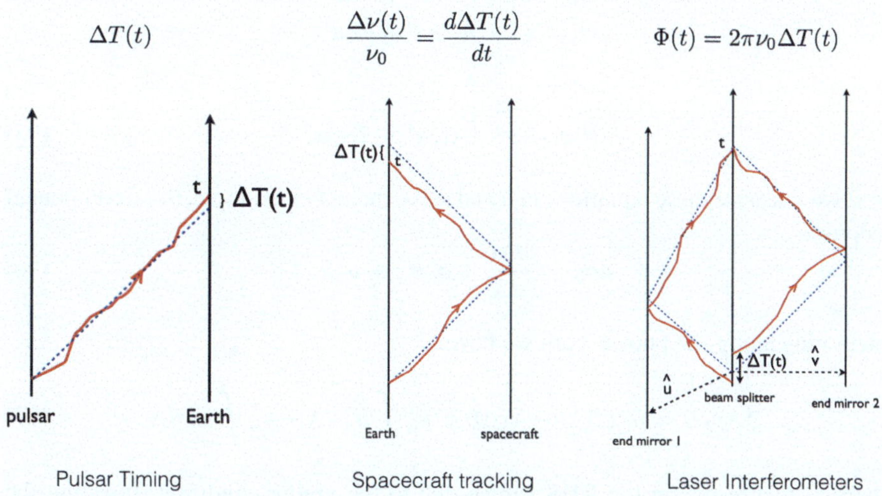

Fig. 11 The measurement principle used by various time-of-flight gravitational wave detectors

$\partial_x, \partial_y, \partial_v$. The associate conserved quantities, combined with the normalization condition $u^\alpha u_\alpha = 0$, fully specifies the four velocity of a photon. If no gravitational waves are present, the photon follows the trajectory $x_0^\mu(\lambda)$ with

$$x_0(\lambda) = x\frac{\lambda}{\Delta\lambda}, \quad y_0(\lambda) = y\frac{\lambda}{\Delta\lambda}, \quad z_0(\lambda) = z\frac{\lambda}{\Delta\lambda}, \quad t_0(\lambda) = \frac{L}{c}\frac{\lambda}{\Delta\lambda} \tag{93}$$

where $L = \sqrt{x^2 + y^2 + z^2}$. The constants of the motion $u_x = \alpha_x, u_y = \alpha_y, u_v = \alpha_v$ in this case are then

$$\begin{aligned}
\alpha_x^0 &= u^x = \frac{dx_0}{d\lambda} = \frac{x}{\Delta\lambda} \\
\alpha_y^0 &= u^y = \frac{dy_0}{d\lambda} = \frac{y}{\Delta\lambda} \\
\alpha_v^0 &= -\frac{1}{2}u^u = -\frac{1}{2}\left(c\frac{dt_0}{d\lambda} - \frac{dz_0}{d\lambda}\right) = \frac{z-L}{2\Delta\lambda}.
\end{aligned} \tag{94}$$

When a gravitational wave is present the photon trajectory is modified: $x^\mu(\lambda) = x_0^\mu(\lambda) + \delta x^\mu(\lambda)$, and we have to adjust the initial direction of the photon to arrive at the same point: $\alpha_\mu = \alpha_\mu^0 + \delta\alpha_\mu$. Both δx^μ and $\delta\alpha_\mu$ are of order h. Note that the TT-gauge "gluing" condition ensures that the spatial coordinate location of the emitter and receiver are unperturbed by the gravitational wave, so that $\delta x^\mu(0) = \delta x^\mu(\Delta\lambda) = 0$.

In the presence of a gravitational wave the constants of motion and the normalization condition become

$$\begin{aligned}
\alpha_x &= (1 + h_+)u^x + h_\times u^y \\
\alpha_y &= (1 - h_+)u^y + h_\times u^x \\
\alpha_v &= -\frac{1}{2}u^u \\
0 &= \alpha_x u^x + \alpha_y u^y + 2\alpha_v u^v .
\end{aligned} \tag{95}$$

Next we linearize these equations in h and solve them. For example, the x component becomes

$$\delta\alpha_x = \frac{d\delta x}{d\lambda} + u_0^x h_+ + u_0^y h_\times . \tag{96}$$

Integrating along the photon path we have

$$\Delta\lambda\,\delta\alpha_x = \delta x(\Delta\lambda) - \delta x(0) + u_0^x \int h_+ d\lambda + u_0^y \int h_\times d\lambda . \tag{97}$$

The first two terms on the RHS vanish due to the gluing condition. Inserting the lowest solution we have

$$\delta\alpha_x = \frac{x\int h_+(u)d\lambda + y\int h_\times(u)d\lambda}{\Delta\lambda^2}. \quad (98)$$

Using the change of variable $du = u^u d\lambda \approx d\lambda(L-z)/\Delta\lambda$ we arrive at the solution

$$\delta\alpha_x = \frac{xH_+ + yH_\times}{(L-z)\Delta\lambda} \quad (99)$$

where $H = \int_0^{L-z} h(u)du$. Repeating the same procedure for the y and v components we have

$$\begin{aligned}\delta\alpha_y &= \frac{xH_+ + yH_\times}{(L-z)\Delta\lambda}\\ \delta\alpha_v &= \frac{c\delta T}{2\Delta\lambda}.\end{aligned} \quad (100)$$

where

$$\delta T = \frac{(x^2-y^2)H_+ + 2xyH_\times}{2cL(L-z)}. \quad (101)$$

The time shift can be cast into coordinate free form by writing the propagation direction as $\hat{k}$ and by writing the vector connecting the two points as $L\hat{a}$ so that

$$\delta T = \frac{(\hat{a}\otimes\hat{a}) : \mathbf{H}}{2c(1-\hat{k}\cdot\hat{a})}. \quad (102)$$

Here $\mathbf{H} = \int_{\xi_1}^{\xi_2} \mathbf{h} d\xi$ is the antiderivative of the gravitational wave tensor

$$\mathbf{h}(\xi) = h_+(\xi)\epsilon^+ + h_\times(\xi)\epsilon^\times, \quad (103)$$

$\xi = ct - \hat{k}\cdot\mathbf{x}_s$, and ϵ^+, $\epsilon^\times$ are the polarization tensors

$$\begin{aligned}\epsilon^+ &= \hat{p}\otimes\hat{p} - \hat{q}\otimes\hat{q}\\ \epsilon^\times &= \hat{p}\otimes\hat{q} + \hat{p}\otimes\hat{q},\end{aligned} \quad (104)$$

and $\hat{p}, \hat{q}$ are vectors that define the principal polarization directions. The colon denotes the double dot product $\mathbf{A} : \mathbf{B} = A_{ij}B^{ij}$.

The signal observed from a gravitational wave source in the θ, ϕ direction $\hat{n} = -\hat{k}$ can be described using the coordinate system shown in Fig. 12 with

$$\begin{aligned}\hat{n} &= \sin\theta\cos\phi\hat{x} + \sin\theta\sin\phi\hat{y} + \cos\theta\hat{z}\\ \hat{u} &= \cos\theta\cos\phi\hat{x} + \cos\theta\sin\phi\hat{y} - \sin\theta\hat{z}\\ \hat{v} &= \sin\theta\hat{x} - \cos\phi\hat{y}.\end{aligned} \quad (105)$$

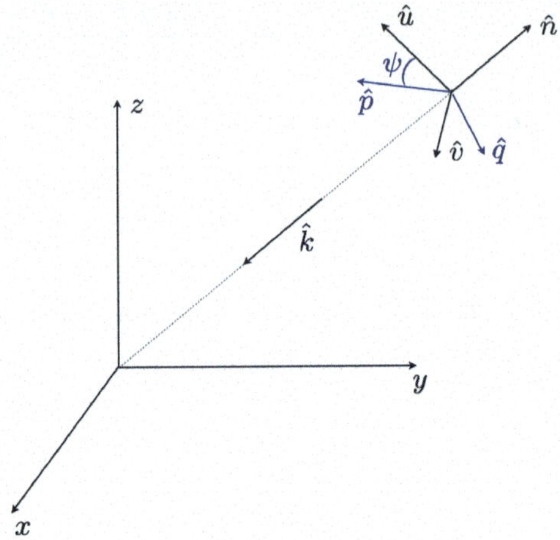

Fig. 12 The coordinate system used to describe the gravitational wave polarization states

The vectors $\hat{u}$, $\hat{v}$ will generally be rotated relative to the principal polarization axes by an angle ψ:

$$\begin{aligned}\hat{p} &= \cos\psi\,\hat{u} + \sin\psi\,\hat{v}\\ \hat{q} &= -\sin\psi\,\hat{u} + \cos\psi\,\hat{v}\,,\end{aligned} \tag{106}$$

so that

$$\begin{aligned}\epsilon^{+} &= \cos(2\psi)\varepsilon^{+} + \sin(2\psi)\varepsilon^{\times}\\ \epsilon^{\times} &= -\sin(2\psi)\varepsilon^{+} + \cos(2\psi)\varepsilon^{\times}\end{aligned} \tag{107}$$

where

$$\begin{aligned}\varepsilon^{+} &= \hat{u}\otimes\hat{u} - \hat{v}\otimes\hat{v}\\ \varepsilon^{\times} &= \hat{u}\otimes\hat{v} + \hat{v}\otimes\hat{u}\,.\end{aligned} \tag{108}$$

As a concrete example, consider a Michelson interferometer. The photon paths we need to consider are shown in Fig. 13. The total time delay is given by

$$\Delta T(t) = \delta T_{12} + \delta T_{24} - \delta T_{13} - \delta T_{34}\,. \tag{109}$$

A kilometer scale detector such as LIGO operating in the frequency band $f \sim 10\,\text{Hz} \to 1000\,\text{Hz}$ has $fL \ll c$, or equivalently $\lambda \gg L$, so we can simplify the calculation by working in the long wavelength limit. Using (102) we find the gravitational wave response to be given by

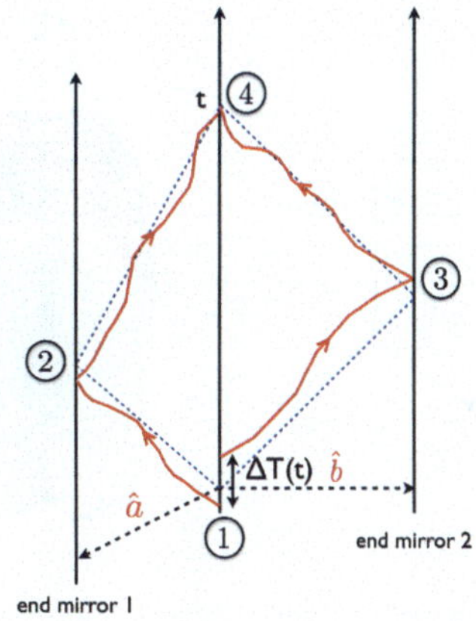

Fig. 13 Photon paths for a Michelson interferometer

$$
\begin{aligned}
h(t) = \frac{c\Delta T(t)}{2L} &= \frac{1}{2}\left[\hat{a}\otimes\hat{a} - \hat{b}\otimes\hat{b}\right] : (h_+(t)\epsilon^+ + h_\times(t)\epsilon^\times) \\
&= F^+h_+(t) + F^\times h_\times(t)\,,
\end{aligned} \tag{110}
$$

where the antenna pattern factors are defined:

$$
\begin{aligned}
F^+ &= \frac{1}{2}\left[\hat{a}\otimes\hat{a} - \hat{b}\otimes\hat{b}\right] : \epsilon^+ \\
F^\times &= \frac{1}{2}\left[\hat{a}\otimes\hat{a} - \hat{b}\otimes\hat{b}\right] : \epsilon^\times\,.
\end{aligned} \tag{111}
$$

If we choose $\hat{a}$ to lie along the x-axis of the coordinate system and $\hat{b}$ to lie along the y-axis, then the various inner products become:

$$
\begin{aligned}
(\hat{a}\otimes\hat{a}) : \varepsilon^+ &= \cos^2\theta\cos^2\phi - \sin^2\phi \\
(\hat{a}\otimes\hat{a}) : \varepsilon^\times &= \cos\theta\sin 2\phi \\
(\hat{b}\otimes\hat{b}) : \varepsilon^+ &= \cos^2\theta\sin^2\phi - \cos^2\phi \\
(\hat{b}\otimes\hat{b}) : \varepsilon^\times &= -\cos\theta\sin 2\phi
\end{aligned} \tag{112}
$$

and (Fig. 14)

$$
\begin{aligned}
F^+ &= \frac{1}{2}(1+\cos^2\theta)\cos(2\phi)\cos(2\psi) - \cos\theta\sin 2\phi\sin 2\psi \\
F^\times &= \frac{1}{2}(1+\cos^2\theta)\cos(2\phi)\sin(2\psi) + \cos\theta\sin 2\phi\cos 2\psi\,.
\end{aligned} \tag{113}
$$

As a second example, consider the time delays measured by pulsar timing. The relevant frequency band is then $f \sim 10^{-9}\,\text{Hz} \rightarrow 10^{-6}\,\text{Hz}$, with the pulsars at $L \sim 1\,\text{kpc}$, so $fL \gg c$ or $\lambda \ll L$. Thus pulsar timing operates in the short wavelength

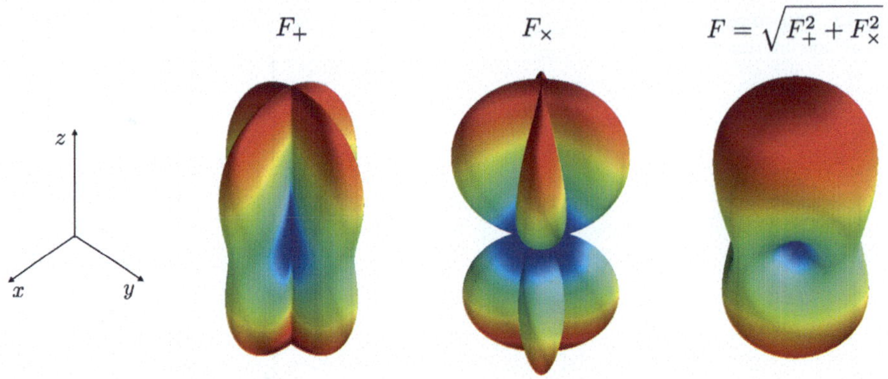

Fig. 14 Antenna patterns in the long wavelength limit for an interferometer with arms along the x and y axes

limit, and we need to consider the full integral along the photon trajectory. For example, a continuous wave with $\mathbf{h} = A\cos(\omega(t - \hat{k}\cdot\mathbf{x}_s))\epsilon^+$ will produce a time-varying time delay from a pulsar in the $\hat{a}$ direction that can be written as

$$\Delta T = \frac{L}{2}\left\{((\hat{a}\otimes\hat{a}):\epsilon^+)\mathrm{sinc}\left[\frac{\omega L}{2c}(1+\hat{k}\cdot\hat{a})\right]\right\}\cos\left[\omega\left(t+\frac{L}{2c}(1+\hat{k}\cdot\hat{a})\right)\right]. \tag{114}$$

We identify the static term in curly brackets to be the wavelength-dependent antenna pattern $F^+(L/\lambda)$, with a similar expression defining the cross polarization. In the pulsar timing literature it is conventional to fix the gravitational wave propagation direction to be $\hat{k}$, and to consider pulsars at different sky locations $\hat{a} \to (\theta, \phi)$. Setting $\hat{k} = -\hat{z}$ we have $\hat{k}\cdot\hat{a} = -\cos\theta$, $(\hat{a}\otimes\hat{a}) : \epsilon^+ = \cos 2\phi \sin^2\theta$ and $(\hat{a}\otimes\hat{a}) : \epsilon^\times = \sin 2\phi \sin^2\theta$. The antenna patterns for the case $L = 10\,\lambda$ are shown in Fig. 15. The time delays are largest for pulsars that are located in roughly the same direction as the source (though the response is zero for pulsars that are in exactly the same direction as the source due to the waves being transverse).

5 Gravitational Wave Observatories

Gravitational wave detection efforts began in the 1950s with Joseph Weber's development of acoustic "bar" detectors. Weber's announcement of a detection in 1969, while ultimately discredited, spurred further interest, even prompting theorists such as Steven Hawking and his student Gary Gibbons to try their hand at gravitational wave detection! The possibility of using laser interferometry as a detection method was proposed in 1963, and the first experimental studies of this approach occurred in 1971. The following year, Rainer Weiss published a landmark study that laid out the

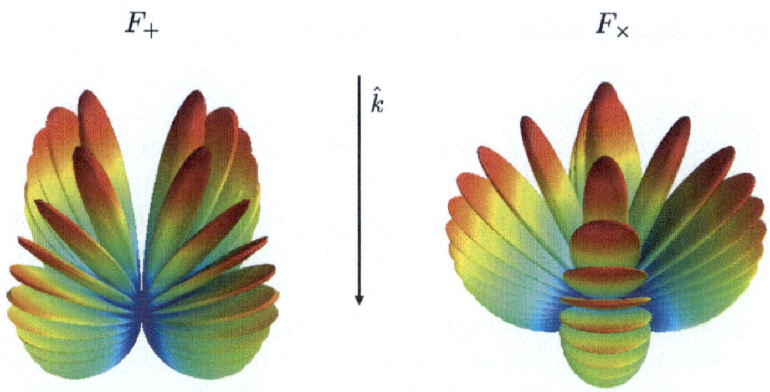

Fig. 15 Antenna patterns as a function of the sky location of the pulsar for a fixed gravitational wave propagation direction $\hat{k}$. The distance to pulsars was set at ten gravitational wavelengths

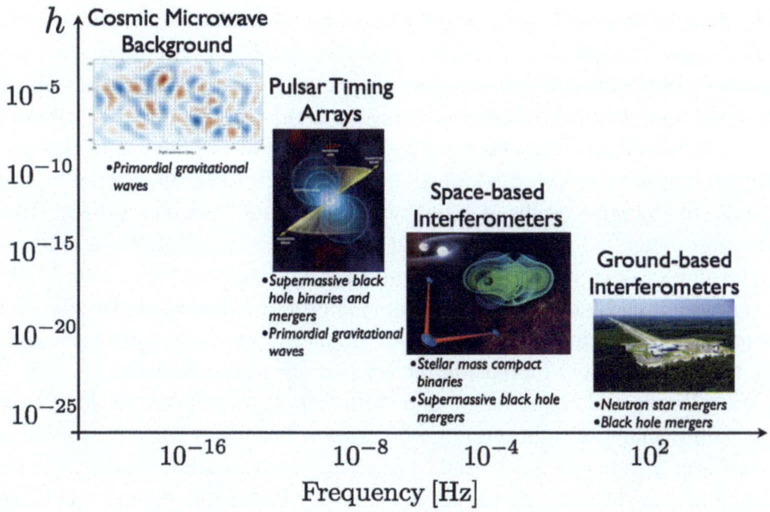

Fig. 16 Gravitational wave detection efforts cover many decades of the gravitational wave spectrum

basic design for a practical laser interferometer gravitational wave detector, paying particular attention to the various sources of noise and how they might be mitigated. The idea of launching a laser interferometer into space appeared a few years later in a report by Weiss, Bender, Pound and Misner. At around the same time, Davies, Anderson, Estabrook and Wahlquist were developing the idea of using spacecraft doppler tracking for gravitational wave detection, which gave Detweiler the idea of using pulsar timing to search for gravitational waves. By the end of the 1970s the basic idea behind the three major detection techniques being pursed today were in place: ground and space based interferometers, and pulsar timing (Fig. 16).

5.1 Ground Based Laser Interferometers

Ground based laser interferometers operate in the audio frequency band $f \sim [10, 10^4]$ Hz, where the primary targets are stellar-remnant mergers of neutron stars and black holes. Other potential sources in the audio band include isolated distorted neutron stars, core-collapse supernovae, low mass X-ray binaries, collapsars and cosmic strings.

The design of a laser interferometer takes into account many factors, but two basic considerations set the overall parameters of the design—maximizing the response to a signal and mitigating laser frequency noise. Interferometers record the time delays due to gravitational waves, $\Delta T(t)$, as phase shifts between laser signals: $\Delta\Phi(t) = 2\pi\nu_0\Delta T(t)$, where ν_0 is the laser frequency. In principle it is possible to make a one-arm interferometer that compares the phase of the laser light returning for a round trip down one arm to a local phase reference. But in practice the laser frequency is not perfectly constant, and frequency fluctuations get multiplied by the overall light travel time to produce phase shifts that would swamp any gravitational wave signals. Using ultra-stable lasers can help mitigate the problem, but even the most stable lasers are still many orders of magnitude too noisy for a one-arm design. The solution is to adopt a Michelson interferometer topology where the laser phase noise is cancelled in the differential arm-length readout. In a Michelson interferometer the input beam is split and sent along two paths of equal length, reflected, then recombined. The laser phase noise is cancelled in the differential read-out. Figure 17 shows a schematic of the interferometer design used for the initial LIGO detectors. The basic Michelson design is augmented by using resonant Fabry–Perot cavities to amplify the signal in each arm. The amplification is achieved by bouncing the laser light back and forth multiple times, effectively increasing the arm-length of the detector.

The overall size and shape of the detectors follow from some simple considerations. As to the shape, a right-angle configuration is chosen since it maximizes the differential arm-length change caused by a quadrupole radiation pattern, as is evident from Fig. 10. As to the size, the longer the arms the larger the response, at least until when the armlength becomes comparable to the gravitational wavelength, a condition which defines the transfer frequency, $f_* = c/(2\pi L)$, where L is the optical path length. For $f < f_*$ we have $\Delta T \sim hL/c$, so to get a large time delay we need a large detector. If the detector is too large we go outside the low frequency limit and the response is diminished. Setting a maximum frequency of 1 kHz defines an optimal size of around 50 km. But building a detector this large would be very costly, so instead resonant cavities are used to fold the light so that the signal gets built up over multiple bounces using much shorter arms—4 km for the LIGO detectors and 3 km for the Virgo detector.

The amplifying effect of a Fabry–Perot cavity can be computed using basic electromagnetic theory. The cavity mirrors can be characterized in terms of their transmissivity and reflectivity, and by keeping track of the electric field transmission and reflection coefficients at each mirror and summing over the multiple bounces yields an expression for the phase shift due to a gravitational wave. In the long wavelength

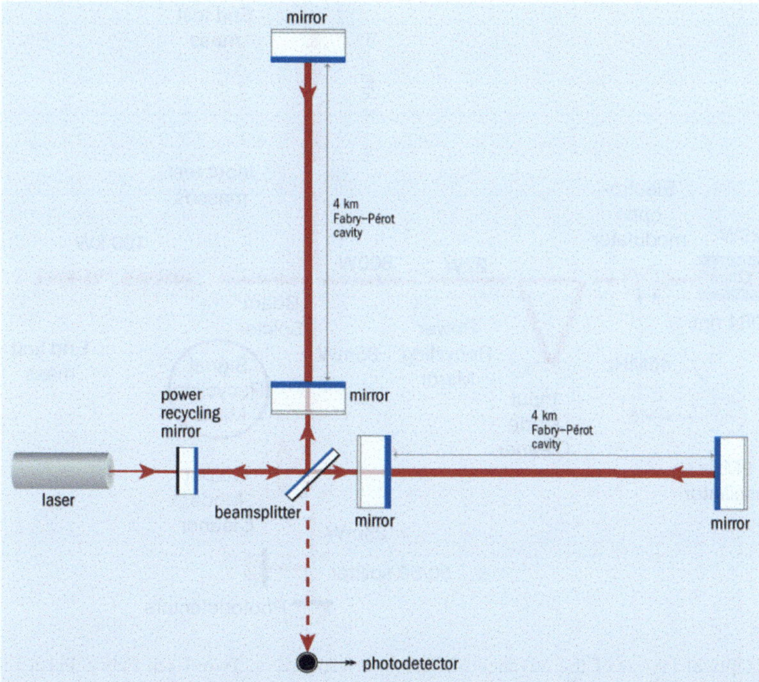

Fig. 17 Basic layout of the initial LIGO interferometers. Two 4-km Fabry–Perot resonant cavities are combined together in a Michelson interferometer topology. Light from the two arms is mixed at the photodetector, yielding a differential measure of the light propagation time in the two arms

limit, and for a gravitational wave that is incident perpendicular to the cavity, the phase shift is given by

$$|\Phi_{\rm FP}| = \left(\frac{\nu_0 h L}{c}\right)\left[\frac{8\mathcal{F}}{\sqrt{1+(f_{\rm gw}/f_p)^2}}\right] \tag{115}$$

where

$$\mathcal{F} = \frac{\pi\sqrt{r_1 r_2}}{1 - r_1 r_2} \tag{116}$$

is the cavity finesse and

$$f_p = \frac{1}{4\pi\tau_s} = \frac{c}{4\mathcal{F}L} \tag{117}$$

is the cavity pole frequency. Here $r_1, r_2 \sim 1$ are the reflectivities of the cavity mirrors, and τ_s is the light storage time in the cavity. Roughly speaking, the cavity boosts the effective length of the arm from L to $2\mathcal{F}L$.

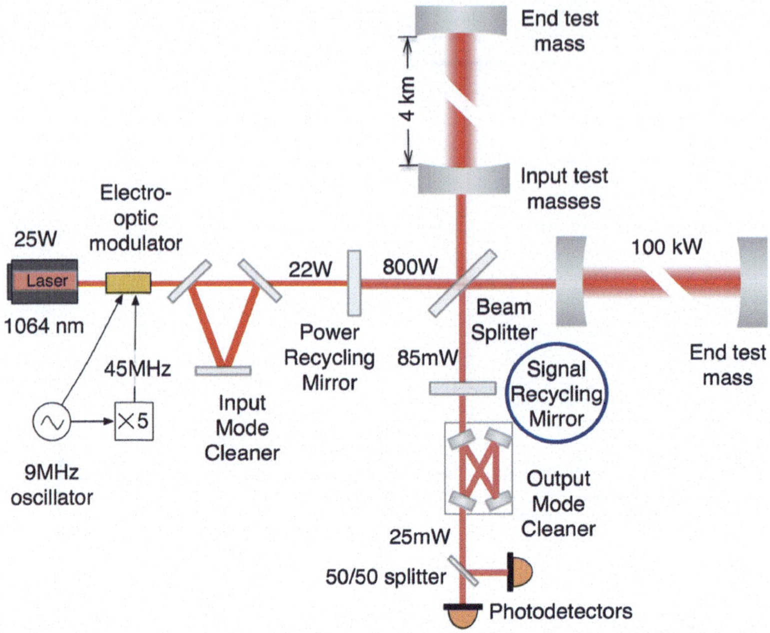

Fig. 18 Optical layout of the advanced LIGO interferometers. Two 4-km Fabry–Perot resonant cavities are coupled together by the signal recycling mirror

To compute the full phase shift due to a gravitational wave for the advanced LIGO and Virgo instruments we need to take into account the fact that the the two Fabry–Perot cavities are now coupled together due to the addition of a signal re-cycling mirror, shown in Fig. 18. The signal recycling mirror allows the cavity to be detuned from resonance in order to recycle the signal at some frequencies and resonantly extract the signal at other frequencies. This makes it possible to manipulate the optical response of the detector as a function of frequency. Without the coupling provided by the signal recycling mirror, the cavity pole frequency for advanced LIGO would sit at the low value of $f_p = 42$ Hz, resulting in poor sensitivity across the band. With the signal recycling mirror the pole frequency for the differential mode is shifted to $f_p \simeq 350$ Hz in the so-called "zero detuned" (fully resonant) configuration, and the signal gain from folding and recycling is roughly a factor of 1,100. The full cavity transfer function for the differential mode is given by the complex function

$$C(f) = \frac{t_s e^{-i(2\pi(f+\nu_0)L_s/c)}}{1 - r_s \left(\frac{r_i - e^{4\pi i f L/c}}{1 - r_i e^{-4\pi i f L/c}}\right) e^{-i(4\pi(f+\nu_0)L_s/c)}} \tag{118}$$

where t_s, r_s, r_i are the transmissivities and reflectivities of the signal recycling mirror and the input mirror, and L_s is the distance to the signal recycling mirror. By changing the phase advance, $\phi = 4\pi L_s/c(f + \nu_0)$ it is possible to tune the response. Examples

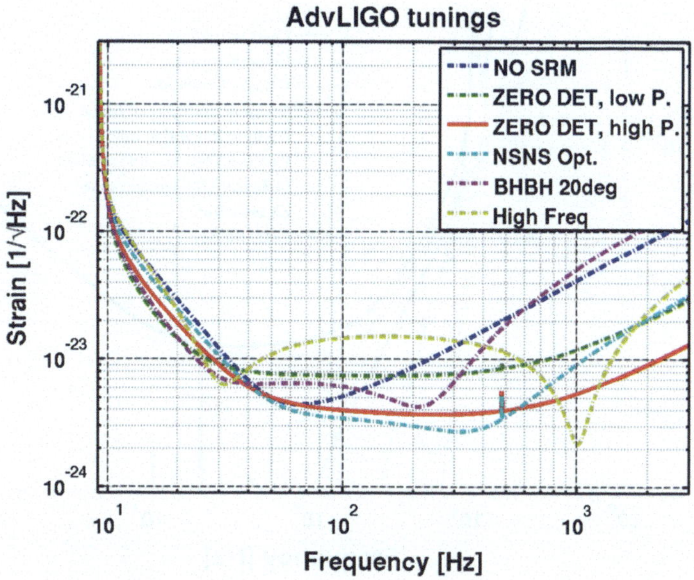

Fig. 19 Examples of the sensitivities that can be achieved by using the placement of the signal recycling mirror to tune the response of the advanced LIGO detectors. The blue line shows the degraded sensitivity that would result if the signal recycling mirror were not used

of how the tunings can modify the instrument sensitivity are shown in Fig. 19. Some of the tunings allow for higher sensitivity in certain narrow frequency bands, which may be useful when targeting particular types of signals.

To understand the sensitivity curves shown in Fig. 19 we need to consider the various noise sources that impact the phase measurement. The noise comes in two flavors, facility noise and fundamental noise. Facility noise includes seismic noise, gravity gradient noise, and various types of thermal noise in the mirrors, mirror coatings and suspensions. Fundamental noise has its origin in quantum mechanics and the Heisenberg uncertainty principle. Figure 20 shows the noise budget—the estimated contribution of each noise term—for the advanced LIGO design.

The overall "U" shape of the noise curve is set by the fundamental quantum sensing noise and the cavity response. The accuracy with which the phase shifts can be measured scales inversely with the number of photons—the more photons the better. This photon counting noise is referred to as "shot noise", and has an amplitude spectral density that scales as $S_{\rm shot}^{1/2} \propto I_0^{-1/2}$, where I_0 is the laser intensity. However, photons carry momentum, and the laser field exerts a fluctuating radiation pressure on the mirrors that scales as $S_{\rm rp}^{1/2} \propto I_0^{1/2}/(Mf^2)$, where M is the mass of the mirror. Thus we have a trade off between minimizing the shot noise by increasing the laser power, versus minimizing the disturbance to the mirrors by lowering the laser power. The cavity response makes the effective laser intensity frequency dependent: $I(f) \simeq I_0/(1 + (f/f_p)^2)$, resulting in a quadrature-sum quantum noise of

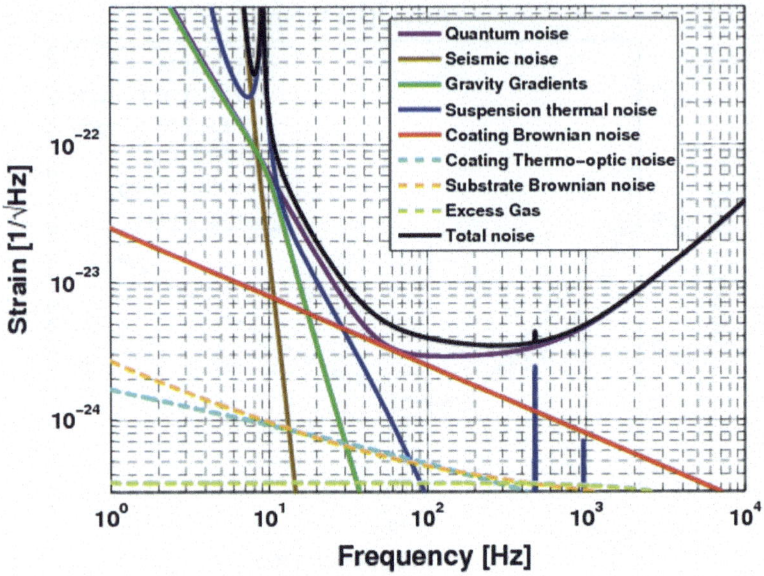

Fig. 20 Noise budget for the advanced LIGO detectors

$$S_Q = \frac{\hbar c}{L^2}\left(\frac{I(f)}{\pi^4 f^4 M^2 c^2} + \frac{1}{I(f)}\right). \tag{119}$$

The frequency at which the two contributions are equal defines what is known as the standard quantum limit. While "quantum limit" sounds rather insurmountable, it turns out not to be. The expression in Eq. (119) assumed that the shot noise and radiation pressure noise are uncorrelated, and so could be added in quadrature, but this does not have to be the case. It is possible to introduce correlations between the terms, and effectively rotating the error ellipse to make the uncertainty in either position or momentum arbitrarily small. This kind of squeezing, or quantum non-demolition measurement was first demonstrated using the GEO600 detector [1].

To get down to the level set by the quantum sensing noise it is necessary to mitigate contributions to the facility noise. Left unattenuated, seismic noise would make gravitational wave detection impossible. The typical level for a ground level facility is of order

$$S_{\rm seis} \simeq 10^{-12}\left(\frac{10\ {\rm Hz}}{f}\right)^2 {\rm Hz}^{-1/2}. \tag{120}$$

The seismic noise can be filtered by suspending the optics on pendula, which introduce transfer functions of the form $|1-(f/f_{\rm pend})^2|$. Setting the pendulum frequency well outside the sensitive band, $f_{\rm pend} \sim 1$ Hz for advanced LIGO, and using a five stage pendulum yields a surpression of $\sim(f_{\rm pend}/f)^{10}$, bringing the seismic noise below the quantum sensing noise across the measurement band.

Thermal noise can be understood in terms of the fluctuation-dissipation theorem, which states that the power spectral density of the fluctuations of a system in equilibrium at temperature T is determined by the dissipative terms that return the system to equilibrium. This can be modeled in terms of an anelastic spring, with dynamics determined by the modified Hooke's law $F = -kx(1 + i\phi)$, where ϕ is the loss angle. The thermal noise is then

$$S_T = \frac{k_B T}{2\pi^3 M f} \frac{f_{\rm res}^2 \phi(f)}{[(f_{\rm res}^2 - f^2)^2 + f_{\rm res}^2 \phi^2(f)]}. \tag{121}$$

Here M is the mirror mass and $f_{\rm res}$ is the resonant frequency of the oscillator. This one expression can be used to estimate mirror coating thermal noise and suspension thermal noise. Suspension thermal noise comes in two forms: pendulum thermal noise and violin mode thermal noise. For the mirror coating thermal noise the resonant frequencies are very high (tens of kHz), and the loss angles very small, and we find

$$S_{\rm MC}^{1/2} \simeq 2.5 \times 10^{-24} \left(\frac{100\,{\rm Hz}}{f}\right)^{1/2} {\rm Hz}^{-1/2}. \tag{122}$$

For the pendulum noise we have $f_{\rm res} = f_{\rm pend} \ll f$ and

$$S_{\rm pend}^{1/2} \simeq 3.5 \times 10^{-25} \left(\frac{100\,{\rm Hz}}{f}\right)^{5/2} {\rm Hz}^{-1/2}. \tag{123}$$

The violin modes described vibrations in the silicon wires used to suspend the optics. The resonant frequencies come in at harmonics of the fundamental mode, which has $f_0 \simeq 500$ Hz. Expanding around the first harmonic we have

$$S_{\rm violin}^{1/2} \simeq \frac{3 \times 10^{-24}}{1 + (f_0^2 - f^2)^2/\delta f^4} {\rm Hz}^{-1/2}, \tag{124}$$

with a linewidth $\delta f = f_0 \phi^{1/2} \simeq 2$ Hz.

Local variations in the gravitational field are an inescapable noise source at low frequencies. For example, a person walking past one of the LIGO end stations exerts a time varying gravitational attraction on the mirrors, which changes the optical path length. A time varying mass distribution $\delta\rho$ generates gravity gradient noise through the gravitational acceleration of the test masses:

$$\ddot{\mathbf{x}} = G \int \frac{\delta\rho(\mathbf{x}', t)}{|\mathbf{x} - \mathbf{x}'|^3} (\mathbf{x} - \mathbf{x}')\, d^3\mathbf{x}'. \tag{125}$$

One of the largest contributions to gravity gradient noise comes from the density fluctuations caused by seismic surface waves. Fluctuations in the density of the atmosphere are also important. These disturbances can be reduced by placing the

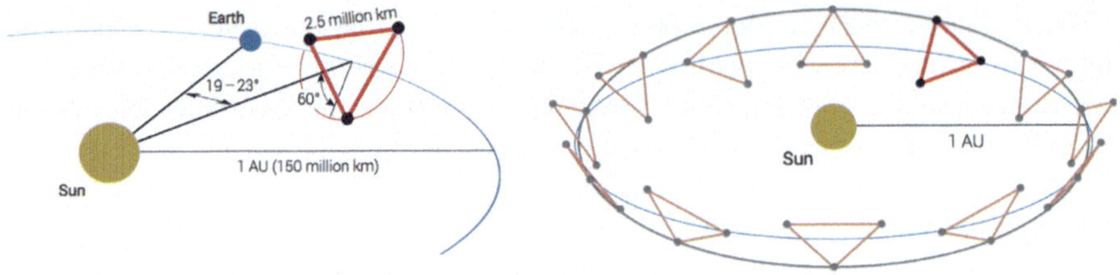

Fig. 21 Proposed orbit for the LISA mission. The constellation cartwheels clockwise as the spacecraft orbit the Sun

detectors deep underground. It may also be possible to use very accurate gravimeters to measure the disturbances and subtract them from the data, but ultimately the best way to escape gravity gradient noise is to put the detectors in deep space.

There is a lot more that could be said about the operation of the LIGO and Virgo detectors that goes well beyond the brief description given here. Topics of particular importance that are not covered here are how the control system keeps the interferometers in resonance, and how the output from this control loop is used to calculate the calibrated strain. To learn more, see Abbott et al. [2], Izumi and Sigg [31].

5.2 *Space Based Laser Interferometers*

To detect signals below $f \sim 1$ Hz we need to get away from gravity gradient and seismic noise. This can be achieved by launching a detector into deep space. The frequency of a gravitational wave signal scales inversely with the size of the source, so space based detectors can detect much larger systems, such as massive black hole mergers and stellar binaries on wide orbits.

The same considerations that guide the design of a ground based interferometer also apply to space detectors, but with the added complication that it is much more difficult to control the distance between the mirrors. Two very different concepts have been proposed for space based interferometers. The first concept, which will be used by the Laser Interferometer Space Antenna [4] (LISA), employs three free-flying spacecraft to form a long baseline detector with synthetic interferometry, while the second concept, which will be used by the Deci-Hertz Interferometer Gravitational wave Observatory [33] (DECIGO), employs precision formation flying, much shorter arms and a resonant Fabry–Perot cavity. The DECIGO concept is essentially LIGO in space, and requires precise control of the spacecraft separations to produce a resonant cavity. The signal would have to be extracted from the control system that maintains the inter-spacecraft separation. In what follows I will focus on the LISA mission since it has already been selected for launch, while DECIGO is at an earlier stage in its development.

The current LISA design calls for three spacecraft to be placed into slightly eccentric and slightly inclined orbits about the Sun, with the center of mass of the constellation trailing the Earth's orbit by about 20°. The orbit is illustrated in Fig. 21. Each spacecraft follows a geodesic Keplerian orbit that is given to leading order in eccentricity by

$$
\begin{aligned}
x_k &= \mathrm{AU}\left(\cos\alpha_k + \frac{e}{2}(\cos(2\alpha_k - \beta_k) - 3\cos\beta_k)\right) \\
y_k &= \mathrm{AU}\left(\sin\alpha_k + \frac{e}{2}(\sin(2\alpha_k - \beta_k) - 3\sin\beta_k)\right) \\
z_k &= -\sqrt{3}e\mathrm{AU}\cos(\alpha_k - \beta_k)
\end{aligned} \tag{126}
$$

where $\alpha_k = 2\pi t/\mathrm{yr} + \kappa$ is the orbit phase of the center of mass and $\beta_k = 2\pi k/3 + \lambda$ is relative phase of the spacecraft in the constellation. The constants κ, λ set the overall orientation and location of the array. To leading order in the eccentricity e the distance between each pair of spacecraft is given by $|\mathbf{x}_k - \mathbf{x}_{k+1}| = L = 2\sqrt{3}\mathrm{AU}e$. The normal to the plane of the constellation makes an angle of 60° with the ecliptic. Setting $L = 2.5 \times 10^6$ km yields an orbital eccentricity of $e = 0.00483$. The armlengths are only constant to leading order in the eccentricity, and even ignoring three body effects from the Earth and other planets, the distance between the spacecraft will change by several meters per second.

The basic idea behind the LISA mission is to use laser interferometry to precisely track the distance between widely separated free flying proof masses. It is necessary to house the proof masses inside a spacecraft to protect them from non-gravitational disturbances such as solar radiation pressure and the solar wind. The spacecraft also provide the platform to house the lasers, optical benches and telescopes for the interferometry system. Figure 22 shows a schematic of the LISA design. The LISA design employs two optical benches on each spacecraft, comprised of a free floating gold/platinum cube and a ~25 cm diameter telescope to transmit and receive the laser signal along each arm. Laser signals are transmitted from each optical bench, producing a total of six laser links. The gold/platinum proof masses are housed in metal cages, and capacitive sensing is used to track the distance between the cubes and the sides of the cage. Micro-Newton trusters gently maneuver the spacecraft to maintain separation between the proof masses and the cages.

The large distances between the spacecraft and the constant changes in the spacecraft separation make it impossible to form a traditional interferometer. For one, the spread in the laser beams over these huge distances mean that very little of the transmitted laser light is received, and even less would be reflected back. Secondly, the changes in the separation amount to millions of phase cycles per second, and even if a Michelson signal could be formed, the differences in the armlengths would lead to the readout being swamped by laser phase noise. The solution is to use synthetic interferometry, where the phase of the incoming laser light is compared to a local reference and recorded by a phasemeter to yield a collection of phase readouts $\phi_{ij}(t)$ that measuring the phase of the signal from spacecraft i that is recorded at spacecraft j at time t. This phase readout will include contributions from laser phase noise C_i,

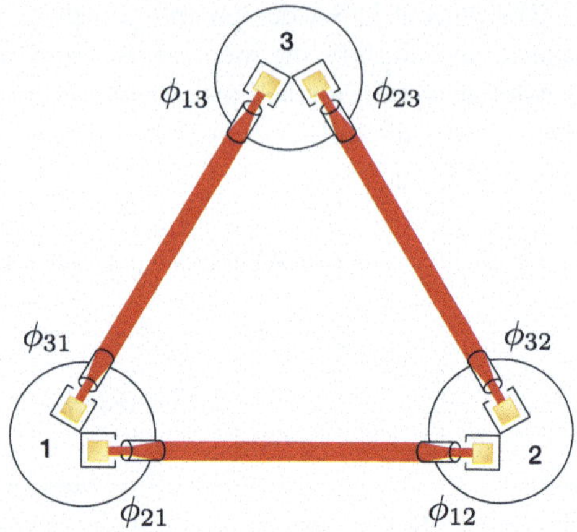

Fig. 22 The LISA measurement system. Laser signals are transmitted from each optical bench, producing a total of six laser links

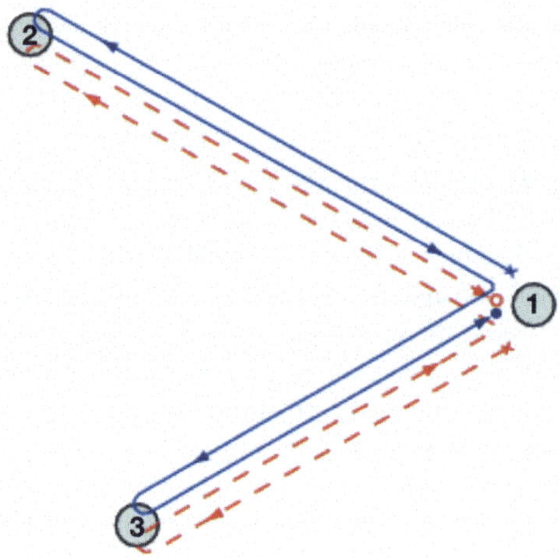

Fig. 23 Synthetic time-delay interferometry works by combing the laser phase measurements along the two virtual paths shown here in red and blue. By going up and back along each arm we ensure that each path has the same total length, and that the total phase difference is therefore free of laser phase noise

position noise n^p_{ij}, acceleration noise $\mathbf{n}^a_{ij}$ and gravitational waves ψ_{ij}:

$$\phi_{ij}(t) = C_i(t - L_{ij}/c) - C_j(t) + n^p_{ij}(t) - \hat{\mathbf{x}}_{ij} \cdot (\mathbf{n}^a_{ij}(t) - \mathbf{n}^a_{ji}(t - L_{ij})) + \psi_{ij}(t)\,, \tag{127}$$

where L_{ij} is the instantaneous length of the arm connecting the two spacecraft and $\hat{\mathbf{x}}_{ij}$ is a unit vector along the arm. The position noise includes a combination of effects in the optical metrology system, the principal one being shot noise in the phase measurement. The acceleration noise is due to non-gravitational forces pushing on the proof masses, for example, due to residual gas in the proof mass housing and feedback from the capacitive sensing and control system. The six phase readouts are combined in software with carefully chosen time delays to synthesize equal-arm length interferometry signals that are free of laser phase noise.

The idea behind synthetic time-delay interferometry is illustrated in Fig. 23. The laser phase noise can be cancelled by following the virtual path shown in Fig. 23 to form the Michelson-like signal:

$$\begin{aligned} X(t) &= \phi_{12}(t - L_{31}/c - L_{13}/c - L_{21}/c) - \phi_{13}(t - L_{21}/c - L_{12}/c - L_{31}/c) \\ &+ \phi_{21}(t - L_{31}/c - L_{13}/c) - \phi_{31}(t - L_{21}/c - L_{12}/c) \\ &+ \phi_{13}(t - L_{31}/c) - \phi_{12}(t - L_{21}/c) + \phi_{31}(t) - \phi_{21}(t)\,. \end{aligned} \tag{128}$$

Similar signals $Y(t)$, $Z(t)$ can be extracted from vertices 2, 3, with expressions that can be found by permuting the indices 1, 2, 3 in the above expression for $X(t)$.

The gravitational wave signal can be computed using the expression given in Eq. (102). The phase shift due to plane gravitational wave propagating in the $-\hat{\Omega}$ direction with frequency f is given by

$$\psi_{ij}(t) = \mathbf{h}(f, t - L_{ij}, \mathbf{x}_i) : (\hat{\mathbf{x}}_{ij} \otimes \hat{\mathbf{x}}_{ij}) \mathcal{T}(\hat{\mathbf{x}}_{ij}, \hat{\Omega}, f) \tag{129}$$

where

$$\mathcal{T}(\hat{\mathbf{x}}_{ij}, \hat{\Omega}, f) = \mathrm{sinc}\left[\frac{f}{2 f_{ij}}(1 - \hat{\mathbf{x}}_{ij} \cdot \hat{\Omega})\right] e^{if/(2f_{ij})(1-\hat{\mathbf{x}}_{ij}\cdot\hat{\Omega})} \tag{130}$$

and $f_{ij} = c/(2\pi L_{ij})$ is the instantaneous transfer frequency. Above the transfer frequency the response diminishes as $\sim 1/f$.

The above expression for the gravitational wave response can be used with the expression for $X(t)$ to compute the sky and polarization averaged response to a gravitational wave. This can be further combined with estimates for the position and acceleration noise to produce a LISA sensitivity curve. The current noise estimates are [27]:

$$P_{\mathrm{pos}} = (1.5 \times 10^{-11}\,\mathrm{m})^2\,\mathrm{Hz}^{-1}\,, \tag{131}$$

and

$$P_{\mathrm{acc}} = (3 \times 10^{-15}\,\mathrm{m\,s}^{-2})^2 \left(1 + \left(\frac{0.4\,\mathrm{mHz}}{f}\right)^2\right) \mathrm{Hz}^{-1}\,. \tag{132}$$

The resulting sensitivity curve, along with some representative LISA signals, is shown in Fig. 24. Details on how the various curves in the figure are calculated can be found in Robson et al. [48].

5.3 *Pulsar Timing Arrays*

Nature has provided us with extremely accurate galactic clocks in the form of pulsars—rapidly rotating Neutron stars which emit beams of radio waves that sweep past the Earth as the Neutron star rotates. The first pulsar, PSR B1919+21, was

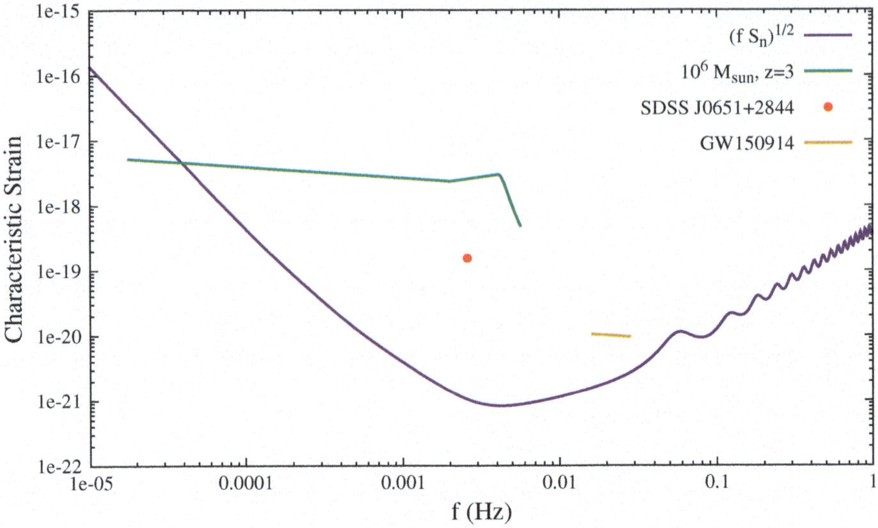

Fig. 24 The LISA sensitivity curve in terms of characteristic strain, $\sqrt{f S_n}$ is compared to three types of signal: an equal mass black hole binary at $z = 3$ with source-frame total mass $M = 10^6\, M_\odot$; the galactic verification binary SDSS J0651+2844 observed for 4 years; and a signal similar to the first LIGO detection GW150914 if the LISA observation started 5 years prior to merger and continued for 4 years

discovered by Jocelyn Bell and Antony Hewish in 1967. The number of pulsars known today is approaching 2,000. Most of these are so-called "classical" pulsars, with spin periods of 0.1–10 s. In 1982 the first millisecond pulsar was discovered [5]. Millisecond pulsars have rotational period in the range of about 1–10 ms, and are thought to be classical pulsars that have been spun-up by accreting material from a binary companion. The accretion is also thought to bury the magnetic field, which reduces the pulsar winds and results in a slower spin down rate. Millisecond pulsars typically have more consistent pulse profiles and more regular pulse periods than classical pulsars, making them much better clocks to use for gravitational wave detection. There are currently 300 known milli-second pulsars, and several new ones are discovered each year.

The idea of using pulsars to detect gravitation waves was first considered by Sazhin in 1978, and in a more general context by Detweiler [17] in 1979, who suggested that cross-correlation of the signals from multiple pulsars could be used to separate noise disturbances from gravitational wave signals. In 1983 Hellings and Downs [26] computed the cross-correlation of the pulse arrival times for pairs of pulsars in the presence of an isotropic stochastic gravitational wave background. They found that the correlation followed the curve

$$C(\theta) = 1 + \frac{3(1-\cos\theta)}{2}\left(\ln\left(\frac{1-\cos\theta}{2}\right) - \frac{1}{6}\right), \tag{133}$$

where θ is the angle between the line of sight to the two pulsars. With N_p pulsars we get $N_p(N_p - 1)/2$ measurements of $C(\theta)$ across a wide range of angles. This correlation pattern is the smoking gun signature that we look for in pulsar timing array searches for gravitational waves. The idea of establishing a pulsar timing array—a collection of millisecond pulsars that are monitored on a regular basis, was first proposed by Forster and Baker in 1990. Since then, three pulsar timing efforts have been undertaken, the Parkes Pulsar Timing Array (PPTA) in Australia, the European Pulsar Timing Array (EPTA) in countries across Europe and the North American NanoHertz Observatory for Gravitational Waves (NANOGrav) in the United States and Canada. The data from all three projects is now analyzed together under the auspices of the International Pulsar Timing Array (IPTA).

Using pulsars to detect gravitational waves sounds simple enough—gravitational waves should perturb the arrival time of the pulses and lead to the distinct correlation pattern predicted by Hellings and Downs—but in practice there are many challenges to overcome. First, both the pulsars and the radio receivers are in constant relative motion, so we have to account for a multitude of effects that contribute to changes in the light propagation time. Another issue is that the individual radio pulses have different shapes, and many thousands of pulses have to be stacked together to arrive at a consistent pulse profile that can be used for the timing. Telescope availability and constraints on observing schedules mean that the timing measurements are irregularly spaced with gaps of one two weeks between observations. The noise levels in each measurement can also vary widely.

The basic equation governing the sensitivity of a radio telescope is the radiometer equation, which leads to the following expression for the noise in the pulse arrival times:

$$\sigma = \left(\frac{S_{\rm psr}}{\rm mJy}\right)^{-1} \left(\frac{T_{\rm rec} + T_{\rm sky}}{\rm Kelvin}\right) \left(\frac{G}{\rm KJy^{-1}}\right)^{-1} \left(\frac{\Delta\nu}{\rm MHz}\right)^{-1/2} \times \left(\frac{t_{\rm int}}{\rm sec}\right)^{-1/2} \left(\frac{W}{P}\right)^{3/2} \left(\frac{P}{\rm ms}\right) \, {\rm ns}\,. \tag{134}$$

Here $S_{\rm psr}$ is the pulsar flux density—brighter pulsars are better; $T_{\rm rec}$, $T_{\rm sky}$ are the system and sky temperature—cool detectors are better; G is the antenna gain—bigger antenna are better; $\Delta\nu$ is the bandwidth—wide band systems are now standard; $t_{\rm int}$ is the integration time, longer is better; W is the pulse width—the narrower the better; and P is the pulse period—the shorter the better. Using the worlds largest radio dishes with wideband receivers and with the brightest millisecond pulsars, it is now possible to measure pulse arrival times with an accuracy of $\sigma \sim 10$ ns.

The pulse arrival times are impacted by many factors. The intrinsic pulse period changes slowly with time, and this can be modeled using a Taylor series expansion for the pulsar period:

$$P(t) = P_0 + \dot{P}_0(t - t_0) + \frac{1}{2}\ddot{P}_0(t - t_0)^2 + \cdots\,. \tag{135}$$

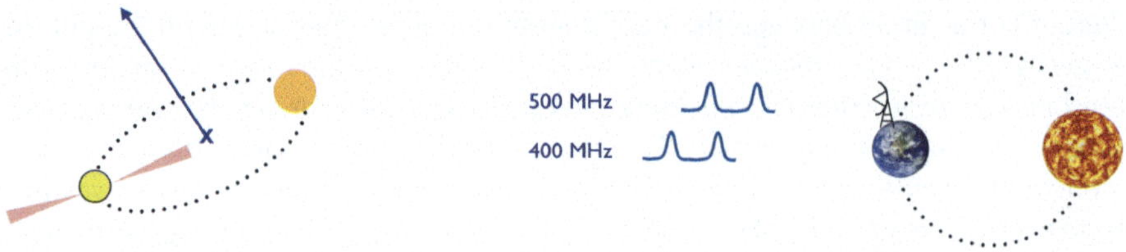

Fig. 25 An illustration of some of the factors that complicate pulsar timing observations. Millisecond pulsars are often found in binary systems that are moving at high velocity through the galaxy. Interstellar dispersion spreads the pulse signals, with the higher frequency radio waves arriving before those at lower frequency. The radio telescopes are on the surface of the Earth, which rotates on its axis and orbits the solar barycenter in a complex orbit that depends on all the other bodies in the solar system

Usually three terms are enough to model the period changes over many decades. As shown in Fig. 25, there are also a host of propagation effects that need to be accounted for. Together with the model for the intrinsic period changes, these propagation effects define the timing model. The predicted pulse arrival time, $t_{\rm pred}$ can be expanded:

$$t_{\rm pred} = t_{\rm PSR} + \Delta_{\odot} + \Delta_{\rm ISM} + \Delta_{\rm B} \,, \tag{136}$$

where $t_{\rm PSR}$ is the emission time at the pulsar, which is modeled using Eq. (135); $\Delta_{\odot}$ maps the arrival time at the telescope to the arrival time at the solar barycenter; $\Delta_{\rm ISM}$ gives the propagation delay from the pulsar system barycenter to the solar barycenter; and $\Delta_{\rm B}$ maps the pulse emission time from the pulsar to the binary barycenter. The difference between the predicted pulse arrival time and the observed pulse arrival time is called a *timing residual*, and these residuals are what are used to search for gravitational waves. Each term in the timing model can be expanded into a collection of contributions. The mapping from the telescope to the solar barycenter can be expanded:

$$\Delta_{\odot} = \Delta_{\rm C} + \Delta_{\rm A} + \Delta_{{\rm R}_\odot} + \Delta_{{\rm E}_\odot} + \Delta_{{\rm S}_\odot} + \cdots \,, \tag{137}$$

where $\Delta_{\rm C}$ are clock corrections; $\Delta_{\rm A}$ are atmospheric delays; $\Delta_{{\rm R}_\odot}$ is the Roemer delay due to the finite speed of light; $\Delta_{{\rm E}_\odot}$ is the Einstein delay due to time dilation for moving clocks and clocks running slower in strong gravitational fields; and $\Delta_{{\rm S}_\odot}$ is the Shapiro delay due to light propagation in a curved spacetime. The Roemer delay depends on the distance between the telescope and the solar barycenter, which is in turn derived from the solar Ephemeris model. Uncertainties in the location of the solar barycenter, which are of order hundreds of meters, have emerged as one of the major sources of uncertainty in the timing model. The mapping from the pulsar to the binary barycenter shares many of the same elements as the solar barycentering:

$$\Delta_{\rm B} = \Delta_{{\rm R}_B} + \Delta_{{\rm E}_B} + \Delta_{{\rm S}_B} + \Delta_{{\rm A}_B} \dots \,. \tag{138}$$

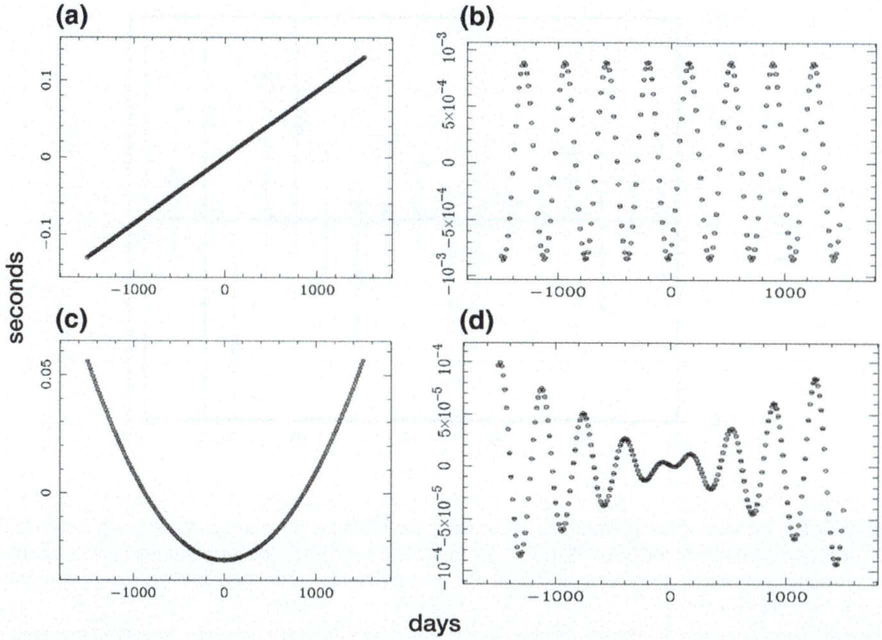

Fig. 26 Examples of timing residuals when there are errors in the timing model. **a** An error in the pulse period P_0; **b** an error in the pulse period derivative $\dot{P}_0$; **c** an error in the sky location, resulting in an error in $\Delta_{\rm VP}$; **d** and error in the pulsar proper motion, resulting in errors in $\Delta_{\rm VP}$ and $\Delta_{\rm E_S}$

Again we have the Roemer delay, $\Delta_{\rm R_B}$; the Einstein delay $\Delta_{\rm E_B}$; and the Shapiro delay $\Delta_{\rm S_B}$; but in addition we have the aberration delay $\Delta_{\rm A_B}$ due to the apparent location of the pulsar being changed by transverse velocity. Once again the Roemer delay includes the full orbital model, which includes relativistic post-Keplerian corrections due to the strong gravitational fields and high velocities encountered in these binaries. Finally there is the propagation delay from the pulsar system to the solar barycenter:

$$\Delta_{\rm ISM} = \Delta_{\rm VP} + \Delta_{\rm ISD} + \Delta_{\rm E_S} + \cdots , \tag{139}$$

where $\Delta_{\rm VP}$ is the vacuum propagation delay—a Roemer type delay that includes the changing distance to the binary; a delay due to interstellar dispersion, $\Delta_{\rm ISD} \sim D/\nu^2$, which scales inversely with the square of the radio frequency; and the Einstein delay $\Delta_{\rm E_S}$ due to the special relativistic time dilation caused by the relative velocity of the binary barycenter. Examples of how some of the terms impact the timing model are shown in Fig. 26. Full details of the timing model can be found in Hobbs et al. [28].

Pulsar timing observations span decades and observations are made $\sim$ weekly, resulting in a sensitivity to gravitational wave signals in the frequency band $f \sim [10^{-9}, 10^{-6}]$ Hz. The main astrophysical source is this band is thought to be supermassive binary black holes with masses in the range $M \sim 10^8 M_\odot - 10^{10} M_\odot$ that

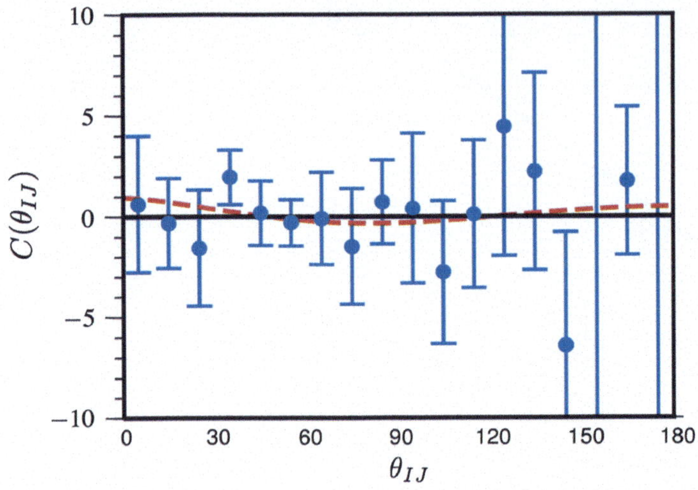

Fig. 27 The current status of attempts to measure the Hellings–Downs correlation curve (dotted red line) using data from the NANOGrav 11 year data release. While the measurement uncertainties are currently very large, they are expected to drop significantly in the next few years, opening the way for a detection

are tens of thousand of years from merger. The combined signal from many hundreds of systems is thought to produce an almost stochastic background that can be detected by cross-correlating the timing residuals from many pulsars. The current status of the search for the correlation pattern by the NANOGrav collaboration [3] is shown in Fig. 27. While the current uncertainties in the correlation measure look to be dauntingly large, simulations suggest that a detection should be possible within the next five years [34].

6 Gravitational Waves from Binary Systems

Binary systems are a prime target for gravitational wave detectors. Binaries made up of ordinary stars are not so promising since they glom together before reaching significant orbital velocities. Compact stellar remnants, such as white dwarfs, Neutron stars and black holes are much better candidates, as are the supermassive black holes found near the centers of galaxies. Binary black hole system are able to reach orbital velocities close to the speed of light before the individual black holes merge to form a larger black hole. The binary dynamics, for both black holes and Neutron stars, is usually divided into three regions: inspiral, merger and ringdown. During the early inspiral the stars follow almost Keplerian orbits, but as the orbit shrinks due to the emission of gravitational waves the orbits become increasingly non-Keplerian, exhibiting interesting relativistic effects such as periastron precession and orbital plane precession. The merger is highly relativistic, and sources strong, dynamical gravitational fields. Neutron star mergers also involve high density material colliding at high velocities. Two black holes merge to form a single distorted black hole, which then sheds the distortions during the ringdown phase. The end state of a Neutron star merger is more complicated, and may involve the formation of a single massive Neutron star that later collapses to form a black hole. Both the massive Neutron star and the final black hole produce ringdown radiation.

In Sect. 3.3 we saw how gravitational waves can be computed in the weak field, slow motion limit. Using more sophisticated techniques, it is possible to continue the weak field expansion of Einstein's equations order-by-order in the orbital velocity v in what is know as the post-Newtonian expansion (PN) of Einstein's equations [45]. The PN expansion breaks down close to merger, and other approximations to Einstein's equations have been developed to cover highly relativistic systems. The self-force program considers the motion of a small compact body with mass m, about a much larger compact body with mass M, and employs an expansion of Einstein's equations in the small mass ratio parameter $q = m/M$ [44, 53]. Another approach is to solve the Einstein equations numerically. Solving Einstein's equations on a computer is a very challenging task, which has required the reformulation of Einstein's equations and the development of many ingenious numerical techniques [35]. The current state-of-the art numerical relativity simulations work well for moderate mass ratios and high velocities, but break down for system with large mass ratios and low velocities. Several schemes have been develop to bridge the gap between the PN approximation and numerical relativity, and to provide analytic waveforms that are valid through inspiral, merger and ringdown. The leading approach recasts the dynamics in terms of an effective one body (EOB) metric, using a transformation similar to that used to solve the Kepler problem in Newtonian gravity [9]. The reformulation improves the convergence properties of the PN expansion. The EOB description is completed by modeling the merger in terms of a single distorted black hole. The current coverage of the binary parameter space, expressed in terms of the orbital velocity v and the mass ratio $q = m/M$, is shown in Fig. 28. For a recent review see Buonanno and Sathyaprakash [10].

In these lectures I will give a brief introduction to the PN expansion, and I will skip the description of the self force program, EOB and numerical relativity. Excellent reviews of these other approaches, along with a far more thorough treatment of the the PN approach, can be found in Poisson [44], Van de Meent [53], Poisson and Will [45], Buonanno and Sathyaprakash [10], Brügmann [8].

6.1 Post-Newtonian Expansion

The PN approach is an expansion of the Einstein field equations in powers of $GM/r \sim v^2$ where M is the total mass of the system, r is the orbital separation and v is the orbital velocity. Several different approaches have been used to compute the PN expansion, including matched asymptotic expansions, solution of the "re-

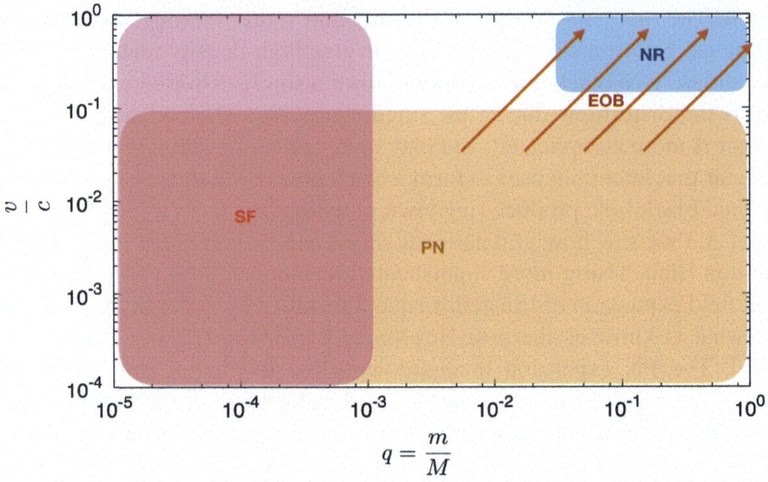

Fig. 28 The regions of the binary system parameter space covered by state-of-the art Post-Newtonian (PN), Self Force (SF) and Numerical Relativity (NR) methods. The Effective One Body (EOB) approach seeks to extend the PN approximation to cover the final plunge, merger and ringdown

laxed" field equations, and effective field theory techniques. For brevity I will skip the derivations and simply quote some of the key results. The relative acceleration of two bodies is expanded:

$$\mathbf{a} = \underbrace{\mathbf{a}_{\rm N}}_{\rm 0PN} + \underbrace{\mathbf{a}_{\rm 1PN}}_{\rm 1PN} + \underbrace{\mathbf{a}_{\rm SO}}_{\rm 1.5PN} + \underbrace{\mathbf{a}_{\rm 2PN}}_{\rm 2PN} + \underbrace{\mathbf{a}_{\rm SS}}_{\rm 2PN} + \underbrace{\mathbf{a}_{\rm RR}}_{\rm 2.5PN} + \cdots . \tag{140}$$

The leading order term is just the usual Newtonian acceleration

$$\mathbf{a}_{\rm N} = -\frac{GM}{r^2}\hat{\mathbf{r}}. \tag{141}$$

Th expressions for the high-order corrections are quite lengthy, so to simplify the notation I will adopt natural units: $G = c = 1$. The first order correction comes in at order v^2, and is responsible for periastron precession:

$$\mathbf{a}_{\rm 1PN} = -\frac{M}{r^2}\left\{\hat{r}\left[(1+3\eta)v^2 - 2(2+\eta)\frac{M}{r} - \frac{3}{2}\eta\dot{r}^2\right] - 2(2-\eta)\dot{r}\mathbf{v}\right\}. \tag{142}$$

Here $\eta = m_1 m_2/M^2 = \mu/M$ is the dimensionless mass ratio, $\mu = m_1 m_2/M$ is the reduced mass and $M = m_1 + m_2$ is the total mass. The next correction to the acceleration enters at order v^3, and is due to spin-orbit coupling

$$\mathbf{a}_{\mathrm{SO}} = \frac{1}{r^3}\left\{6\dot{r}\left[(\hat{r}\times\mathbf{v})\cdot\left(2\mathbf{S}+\frac{\delta m}{M}\boldsymbol{\Delta}\right)\right]-\left[\mathbf{v}\times\left(7\mathbf{S}+3\frac{\delta m}{M}\boldsymbol{\Delta}\right)\right]\right.$$
$$\left.+3\dot{r}\left[\hat{r}\times\left(3\mathbf{S}+\frac{\delta m}{M}\boldsymbol{\Delta}\right)\right]\right\}. \tag{143}$$

Here $\delta m \equiv m_1 - m_2$ is the mass difference, $\mathbf{S} \equiv \mathbf{S_1} + \mathbf{S_2}$ is the total spin and $\boldsymbol{\Delta} \equiv M(\mathbf{S_2}/m_2 - \mathbf{S}_1/m_1)$. Two effects enter at order v^4. The first is due to gravitational self interaction (gravity gravitates)

$$\mathbf{a}_{\mathrm{2PN}} = -\frac{M}{r^2}\left\{\hat{r}\left[\frac{3}{4}(12+29\eta)\left(\frac{M}{r}\right)^2+\eta(3-4\eta)v^4+\frac{15}{8}\eta(1-3\eta)\dot{r}^4\right.\right.$$
$$\left.-\frac{3}{2}\eta(3-4\eta)v^2\dot{r}^2-\frac{1}{2}\eta(13-4\eta)\frac{M}{r}v^2-(2+25\eta+2\eta^2)\frac{M}{r}\dot{r}^2\right]$$
$$\left.-\frac{1}{2}\dot{r}\mathbf{v}\left[\eta(15+4\eta)v^2-(4+41\eta+8\eta^2)\frac{M}{r}-3\eta(3+2\eta)\dot{r}^2\right]\right\}, \tag{144}$$

while the second is due to spin-spin coupling:

$$\mathbf{a}_{\mathrm{SS}} = -\frac{3}{\mu r^4}\left\{\hat{r}(\mathbf{S}_1\cdot\mathbf{S}_2)+\mathbf{S}_1(\hat{r}\cdot\mathbf{S}_2)+\mathbf{S}_2(\hat{r}\cdot\mathbf{S}_1)-5\hat{r}(\hat{r}\cdot\mathbf{S}_1)(\hat{r}\cdot\mathbf{S}_2)\right\}. \tag{145}$$

All the terms considered so far are time reversal invariant and do not cause the energy and angular momentum of the orbit to evolve. The first non-time-reversal-invariant term arises at order v^5 and is due to radiation reaction:

$$\mathbf{a}_{\mathrm{RR}} = \frac{8}{5}\eta\frac{M^2}{r^3}\left\{\dot{r}\hat{\mathbf{r}}\left[18v^2+\frac{2}{3}\frac{M}{r}-25\dot{r}^2\right]-\mathbf{v}\left[6v^2-2\frac{M}{r}-15\dot{r}^2\right]\right\}. \tag{146}$$

The radiation reaction term causes the orbit to decay. To see this, consider a circular orbit with $\dot{r} = 0$ and $v^2 = M/r$. The torque due to the radiation reaction force is given by

$$\tau_{RR} = \mu\mathbf{r}\times\mathbf{a}_{\mathrm{RR}} = -\frac{32}{5}\frac{\eta}{M}v^8\mathbf{L}_{\mathrm{N}} \tag{147}$$

where $\mathbf{L}_{\mathrm{N}} = \mu\mathbf{r}\times\mathbf{v}$ is the Newtonian orbital angular momentum. Because the radiation reaction torque is directed against the orbital angular momentum, it causes the orbit to decay.

In addition to specifying the relative acceleration of the two masses, the PN equations of motion also describe the evolution of the spins of the two bodies, which evolve due to relativistic spin-orbit and spin-spin interactions. Up to 2 PN order the total angular momentum $\mathbf{J} = \mathbf{L} + \mathbf{S}_1 + \mathbf{S}_2$ is conserved, and the spins and orbital angular momentum obey the precession equation

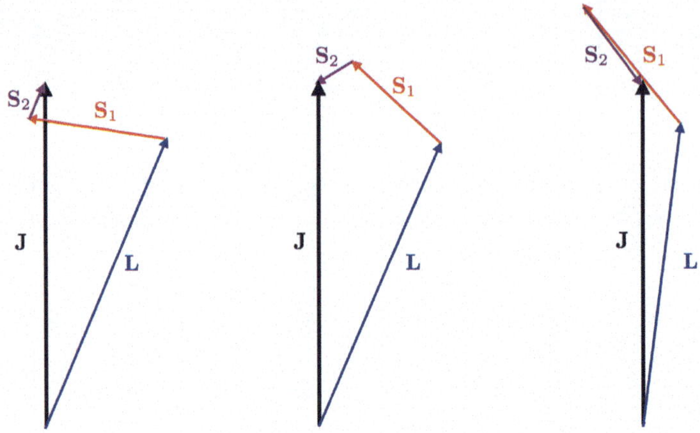

Fig. 29 Snapshots showing an example of how the individual spins $\mathbf{S}_1$, $\mathbf{S}_2$ and the orbital angular momentum $\mathbf{L}$ evolve due to spin-orbit and spin-spin interactions at 2 PN order

$$
\begin{aligned}
\dot{\mathbf{L}}_{\rm N} &= -\frac{1}{r^3}\left\{\left[\mathbf{L}_{\rm N} \times \left(\frac{7}{2}\mathbf{S} + \frac{3}{2}\frac{\delta m}{m}\boldsymbol{\Delta}\right)\right] + 3(\hat{r}\cdot\mathbf{S}_1)(\hat{r}\times\mathbf{S}_2) + 3(\hat{r}\cdot\mathbf{S}_2)(\hat{r}\times\mathbf{S}_1)\right\}, \\
\dot{\mathbf{S}}_1 &= \frac{1}{r^3}\left\{(\mathbf{L}_{\rm N}\times\mathbf{S}_1)\left(2+\frac{3}{2}\frac{m_1}{m_2}\right) - \mathbf{S}_2\times\mathbf{S}_1 + 3(\hat{r}\cdot\mathbf{S}_2)\hat{\mathbf{r}}\times\mathbf{S}_1\right\}, \\
\dot{\mathbf{S}}_2 &= \frac{1}{r^3}\left\{(\mathbf{L}_{\rm N}\times\mathbf{S}_2)\left(2+\frac{3}{2}\frac{m_1}{m_2}\right) - \mathbf{S}_1\times\mathbf{S}_2 + 3(\hat{r}\cdot\mathbf{S}_1)\hat{\mathbf{r}}\times\mathbf{S}_2\right\}.
\end{aligned}
\tag{148}
$$

Figure 29 shows snapshots of how the individual spins $\mathbf{S}_1$, $\mathbf{S}_2$ and the orbital angular momentum $\mathbf{L}$ evolve due to spin-orbit and spin-spin interactions at 2 PN order. At this order there is no dissipation and the magnitude of the spins and the orbital angular momentum are constant. The total angular momentum $\mathbf{J}$ remains constant at 2 PN order. When dissipation is included the magnitude of $\mathbf{L}$ and $\mathbf{J}$ both decrease, but the orientation of the total angular moment $\hat{J}$ and the magnitude of the spins remain constant to a high degree of accuracy. A complete analytic solution to the 2 PN order (non-dissipative) spin precession equations was only found quite recently. The motion can be described using Jacobi elliptic functions. The solution has since been extended to include 2.5 PN order dissipative effects.

6.2 *Circular Newtonian Binary*

It is instructive to consider the gravitational waves generated by the simplest binary system imaginable—a circular binary at Newtonian order. Writing locations of the two masses as $\mathbf{x}_1$ and $\mathbf{x}_2$, the motion can be described using center-of-mass $\mathbf{x}_{\rm COM} = (m_1\mathbf{x}_1 + m_2\mathbf{x}_2)/M$ and relative coordinates $\mathbf{x} = \mathbf{x}_1 - \mathbf{x}_2 = r\,\hat{r}$. At leading Newtonian order $\ddot{\mathbf{x}}_{\rm COM} = 0$ and $\ddot{\mathbf{x}} = -(M/r^2)\hat{r}$. We can choose coordinates where the orbital motion is restricted to the x, y plane and (Fig. 30)

$$\mathbf{x} = -r\sin(\omega t)\hat{x} + r\cos(\omega t)\hat{y} \tag{149}$$

where the equations of motion demand $\omega^2 r^3 = M$. The individual masses follow the orbits

$$\mathbf{x}_1 = \frac{m_2}{M}\left(-r\sin(\omega t)\hat{x} + r\cos(\omega t)\hat{y}\right), \quad \mathbf{x}_2 = \frac{m_1}{M}\left(r\sin(\omega t)\hat{x} - r\cos(\omega t)\hat{y}\right). \tag{150}$$

The mass quadrupole moment tensor is given by

$$Q^{ij} = \int d^3x\, \rho(t,\mathbf{x})\left(x^i x^j - \frac{1}{3}r^2\delta^{ij}\right) \tag{151}$$

with mass density

$$\rho(t,\mathbf{x}) = m_1\delta(\mathbf{x}-\mathbf{x}_1(t)) + m_2\delta(\mathbf{x}-\mathbf{x}_2(t)). \tag{152}$$

The non-vanishing components of the mass quadrupole tensor are then given by

$$\begin{aligned} Q^{xx} &= \mu r^2\left(\sin^2(\omega t) - \frac{1}{3}\right), \quad Q^{yy} = \mu r^2\left(\cos^2(\omega t) - \frac{1}{3}\right), \\ Q^{xy} &= -\mu r^2\sin(2\omega t), \quad Q^{zz} = -\frac{1}{3}\mu r^2. \end{aligned} \tag{153}$$

To compute the gravitational wave strain we need the second time derivatives of these quantities, which are given by

$$\begin{aligned} \ddot{Q}^{xx} &= \frac{2\mu M}{r}\cos(2\omega t), & \ddot{Q}^{yy} &= -\frac{2\mu M}{r}\cos(2\omega t) \\ \ddot{Q}^{xy} &= \frac{4\mu M}{r}\sin(2\omega t), & \ddot{Q}^{zz} &= 0. \end{aligned} \tag{154}$$

Using the coordinate system defined in panel (a) of Fig. 30 and recalling Eq. (105), we find that the TT waveform $h_{ij}^{\rm TT} = \frac{2}{R}P_{ijkl}\ddot{Q}^{kl}(t-r)$ has non-vanishing components given by

$$\begin{aligned} h_{uu} &= -h_{vv} = h_+ = \frac{2\mu M}{rR}(1+\cos^2\theta)\cos(2\omega t + 2\phi), \\ h_{uv} &= h_{vu} = h_\times = \frac{4\mu M}{rR}\cos\theta\sin(2\omega t + 2\phi). \end{aligned} \tag{155}$$

For an orbit with a general orientation, as shown in panel (b) of Fig. 30 the waveform is given by

$$\begin{aligned} h_{uu} &= -h_{vv} = \cos(2\psi)h_+ + \sin(2\psi)h_\times, \\ h_{uv} &= h_{vu} = -\sin(2\psi)h_+ + \cos(2\psi)h_\times, \end{aligned} \tag{156}$$

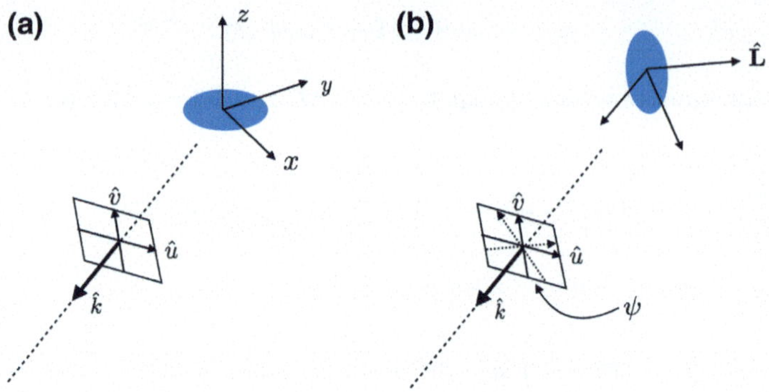

Fig. 30 The coordinate systems we are using to describe a circular, Newtonian binary. Panel **a** is for a binary where the orbital angular momentum is aligned with the z axis of the coordinate system, while panel **b** is for the case where the orbit has a general orientation

and

$$h_+ = \frac{2\mu M}{rR}(1+\cos^2 \iota)\cos(2\omega t),$$
$$h_\times = \frac{4\mu M}{rR}\cos\iota\sin(2\omega t)\,. \tag{157}$$

where ι is the orbital inclination with respect to the line of sight to the binary and ψ is the polarization angle, which are given by

$$\cos\iota = \hat{n}\cdot\hat{L} = -\hat{k}\cdot\hat{L},$$
$$\tan\psi = \frac{\hat{v}\cdot(\hat{n}\times\hat{L})}{\hat{u}\cdot(\hat{n}\times\hat{L})}\,. \tag{158}$$

Using the Kepler condition $\omega^2 r^3 = M$ we can re-express the gravitational wave amplitude in terms of the orbital angular frequency:

$$h_+ = \frac{2\mathcal{M}^{5/3}\omega^{2/3}}{R}(1+\cos^2 \iota)\cos(2\omega t),$$
$$h_\times = \frac{4\mathcal{M}^{5/3}\omega^{2/3}}{R}\cos\iota\sin(2\omega t)\,. \tag{159}$$

where $\mathcal{M} = (m_1 m_2)^{3/5}/M^{1/5}$ is the *chirp mass*. Using Eq. (88) we can compute the energy radiated per unit solid angle averaged over a wave cycle:

$$\left(\frac{dP}{d\Omega}\right)_{\rm quad} = \frac{R^2}{16\pi}\langle \dot{h}_+^2 + \dot{h}_\times^2\rangle = \frac{2\mu^2 r^4\omega^6}{\pi}\left[\left(\frac{1+\cos^2\iota}{2}\right)^2 + \cos^2\iota\right]. \tag{160}$$

Note that the emission depends on the inclination ι, but not on the polarization angle ψ. Integrating over ι, ψ we find that the total energy flux is equal to

$$P_{\text{quad}} = \frac{32\mu^2 r^4 \omega^6}{5} = \frac{32}{5}(\mathcal{M}\omega)^{10/3} = \frac{32}{5}\eta^2 v^{10}. \tag{161}$$

We can now use an energy balance argument to compute the back-reaction on the orbit. The orbital energy in the Newtonian limit is given by

$$E = \frac{1}{2}\mu v^2 - \frac{\mu M}{r} = -\frac{\mu M}{2r} = -\frac{1}{2}\mathcal{M}^{5/3}\omega^{2/3}. \tag{162}$$

Setting $dE/dt = -P_{\text{quad}}$ yields the balance equation

$$\dot{\omega} = \frac{96}{5}\mathcal{M}^{5/3}\omega^{11/3}. \tag{163}$$

Integrating the balance equation we find that the orbital frequency will grow with time:

$$\omega(t) = \frac{1}{\mathcal{M}}\left(\frac{5\mathcal{M}}{256(t_c - t)}\right)^{3/8}. \tag{164}$$

Here t_c is a reference time where the orbital frequency formally becomes infinite. In reality the black holes or Neutron stars will merge before this point, and the divergence in the frequency indicates a break down of the PN description. Note that the amplitude of the signal is proportional to $\omega^{2/3}$, so both the frequency and amplitude increase with time. Signals that increase in volume and pitch are called chirps, and since the combination of masses $\mathcal{M}$ sets the timescale for the evolution it is called the chirp mass. Higher order terms in the PN expansion modify the expression for the chirp (164), but even the leading order expression does a very good job of modeling the time evolution of a neutron star merger, as can been seen in the comparison between the binary neutron start signal GW170817 seen in the LIGO Livingston detector and leading order prediction for the frequency evolution with time shown in Fig. 31.

The orbital frequency evolution (164) can be integrated with respect to time to give the orbital phase evolution:

$$\Phi(t) = \int \omega(t)dt = \Phi_c - \frac{1}{32}\left(\frac{256(t_c - t)}{5\mathcal{M}}\right)^{5/8}, \tag{165}$$

and the hence the full chirp signal

$$h(t) = \frac{4\mathcal{M}}{R}\left(\frac{5\mathcal{M}}{256(t_c - t)}\right)^{1/4} \cos 2\Phi(t). \tag{166}$$

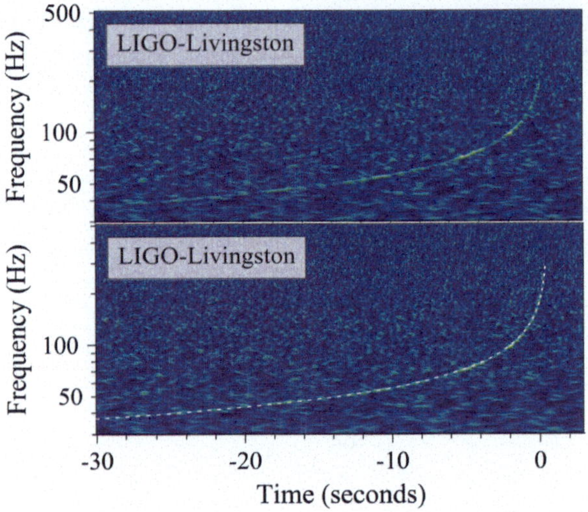

Fig. 31 A whitened spectrogram of the LIGO Livingston data showing the Neutron star merger GW170817. The yellow line visible in the upper panel shows the frequency evolution of the observed signal, while the dashed white line overlaying signal in the lower panel was generated using Eq. (164) with chirp mass $\mathcal{M} = 1.19 M_\odot$

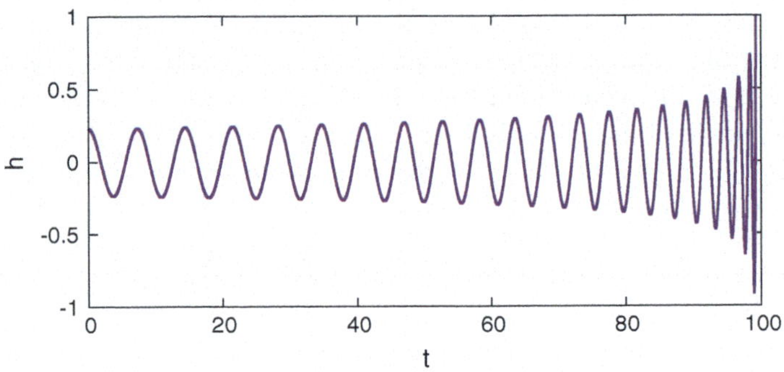

Fig. 32 An example of a leading PN order chirp waveform

An example of a chirp waveform is show in Fig. 32.

6.3 *Stationary Phase Approximation*

PN waveforms are computed in the time domain, but in many cases gravitational wave data analysis is carried out in the frequency domain. The time dominated waveforms can be sampled and transformed to the frequency domain via a Fast Fourier Transform (FFT), but it is more efficient to analytically transform the waveforms using the stationary phase approximation (SPA). The SPA works very well for a wide range of PN waveforms.

The Fourier transform of a signal is defined:

$$\tilde{h}(f) = \int_{-\infty}^{\infty} h(t) e^{2\pi i f t}\, dt = \int_{-\infty}^{\infty} A(t) e^{-2\pi i \varphi(t)} e^{2\pi i f t}\, dt\,, \tag{167}$$

where we have written $h(t)$ in terms of the time dependent amplitude $A(t)$ and gravitational wave phase $\varphi(t)$. The SPA is computed at the stationary time t_* where $\dot{\varphi}(t_*) = 2\pi f$. Taylor expanding the phase about this point we have

$$\varphi(t) = \varphi(t_*) + 2\pi f (t - t_*) + \frac{1}{2}\ddot{\varphi}(t_*)(t - t_*)^2 + \cdots \tag{168}$$

The amplitude is assumed to be slowly varying and can be treated as constant near the stationary point (Fig. 33). The Fourier integral becomes

$$\begin{aligned} \tilde{h}(f) &\simeq A(t_*) e^{i(2\pi f t_* - \varphi(t_*))} \int_{-\infty}^{\infty} e^{-\frac{i}{2}\ddot{\varphi}(t_*)(t-t_*)^2}\, dt \\ &= A(t_*) e^{i(2\pi f t_* - \varphi(t_*) - \pi/4)} \left(\frac{2\pi}{|\ddot{\varphi}(t_*)|}\right)^{1/2}. \end{aligned} \tag{169}$$

The SPA breaks down if $A(t)$ varies too rapidly, or if $\ddot{\varphi}(t_*)$ vanishes at the stationary point. The later can occur for systems with spin precession. Both conditions are violated for black hole ringdowns. Applying the SPA to the leading order PN waveform

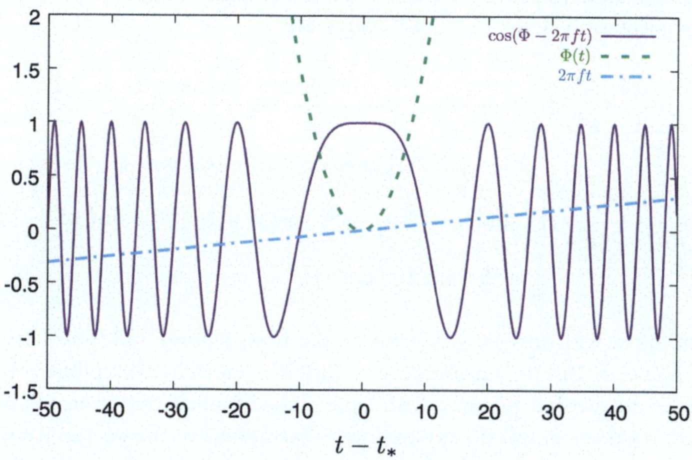

Fig. 33 A graphical illustration of why the stationary phase approximation works. Near the stationary point t_* the integrand is slowly varying and gives a non-zero contribution. Away from the stationary point the integrand oscillates rapidly and averages to zero

(165) yields

$$\tilde{h}(f) \simeq \left(\frac{5}{6}\right)^{1/2} \frac{\mathcal{M}^{5/6}}{R\pi^{2/3}} f^{-7/6} e^{i(2\pi f t_c - \phi_c - \pi/4 + 3/4(8\mathcal{M} f)^{-5/3})} . \tag{170}$$

6.4 Eccentric Newtonian Binary

Orbital eccentricity introduces several interesting effects in the waveforms. At Newtonian order the orbital motion is given by the Kepler solution for the orbital phase ϕ and radial separation r:

$$\begin{aligned} \phi &= \phi_0 + \arctan\left(\left(\frac{1+e}{1-e}\right)^{1/2} \tan\frac{u}{2}\right) \\ r &= a(1 - e\cos u) \\ \omega t &= u - e\sin u . \end{aligned} \tag{171}$$

The orbital angular frequency ω and the semi-major axis a are related by the Kepler equation: $\omega^2 a^3 = M$. The orbital eccentricity can be expressed in terms of the energy and angular moment as

$$e^2 = 1 + \frac{2EL^2}{M^2\mu^2} . \tag{172}$$

The gravitational waveforms can be computed using the same steps as for the circular case, though the algebra gets considerably more involved. The final result for the plus and cross polarization states in the TT gauge are:

$$\begin{aligned} h_+ &= \frac{\mu M}{Ra(1-e^2)}\left(\left[2\cos(2\phi + 2\psi) + \frac{5e}{2}\cos(\phi + 2\psi) + \frac{e}{2}\cos(3\phi + 2\psi)\right.\right. \\ &\qquad \left.\left. + e^2\cos 2\psi\right](1 + \cos^2\iota) + [e\cos\phi + e^2]\sin\iota\right) \\ h_\times &= \frac{\mu M}{Ra(1-e^2)}\left([4\sin(2\phi + 2\psi) + 5e\sin(\phi + 2\psi) + e\sin(3\phi + 2\psi)\right. \\ &\qquad \left. + 2e^2\sin 2\psi]\cos\iota\right) . \end{aligned} \tag{173}$$

We see that the waveforms now depend on the first, second and third harmonics of the orbital phase ϕ. But the harmonic structure is even richer than this as the orbital phase ϕ does not evolve linearly with time. Indeed, each harmonic of the orbital frequency introduces an infinite collection of harmonics of the orbital frequency via the relation

$$\cos\phi = -e + \frac{2(1-e^2)}{e}\sum_{k=1}^{\infty} J_k(ke)\cos(k\omega t) , \tag{174}$$

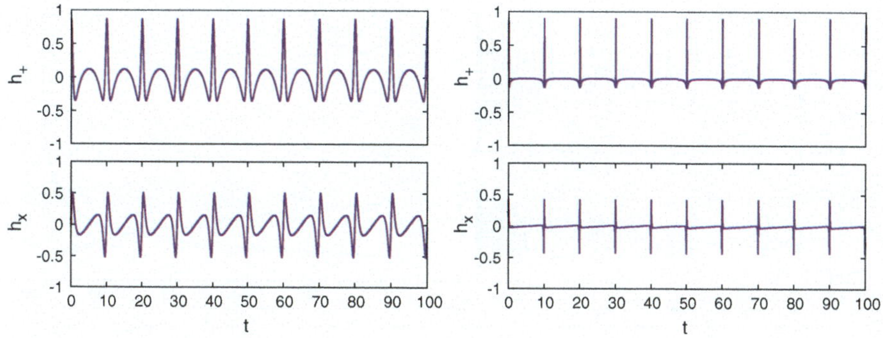

Fig. 34 Plus and cross waveforms for eccentric binaries at Newtonian order. The panel on the left has $e = 0.5$ while the panel on the right has $e = 0.9$. Systems will large eccentricities produce a burst of radiation at periapse

where J_k are Bessel functions of the first kind. In the limit $e \to 0$ only the $k = 1$ term survives. Conversely, in the limit $e \to 1$ it no longer make sense to describe the waveform in terms of individual harmonics, and we find instead that the signal is better described in terms of discrete bursts of radiation at periapse. Figure 34 shows the plus and cross polarizations for systems with $e = 0.5$ and $e = 0.9$ viewed at an inclination angle of $\iota = 60°$.

The emission of gravitational waves takes away energy and angular momentum from the system. The energy and angular momentum loss can be computed using Eqs. (89) and (91):

$$\frac{dE}{dt} = \frac{96\mu^2 M^3}{5a^5} \frac{1}{(1-e^2)^{7/2}} \left(1 + \frac{73}{24}e^2 + \frac{37}{96}e^4\right)$$
$$\frac{dL}{dt} = -\frac{32\mu^2 M^{5/2}}{5a^{7/2}} \frac{1}{(1-e^2)^2} \left(1 + \frac{7}{8}e^2\right) . \tag{175}$$

Combining these equations with $E = -M\mu/(2a)$, $L^2 = \mu^2 Ma(1-e^2)$ and $\omega^2 a^3 = M$ yields adiabatic evolution equations for the orbital frequency and the eccentricity:

$$\frac{d\omega}{dt} = \frac{32\mathcal{M}^{5/3}\omega^{11/3}}{5} \frac{1}{(1-e^2)^{7/2}} \left(1 + \frac{73}{24}e^2 + \frac{37}{96}e^4\right)$$
$$\frac{de}{dt} = -\frac{304\mathcal{M}^{5/3}\omega^{8/3}}{15} \frac{e}{(1-e^2)^{5/2}} \left(1 + \frac{121}{304}e^2\right) . \tag{176}$$

If we drop terms of order e^2 and higher the system can be recast as $d\ln e/d\ln\omega \approx -19/18$, which tells us that systems loose roughly a decade in eccentricity for every decade in frequency. Typical binary systems are thought to form with low velocities and low orbital frequencies. Even if the binaries are initially very eccentric, the emission of gravitational waves will circularize the system before merger. There are

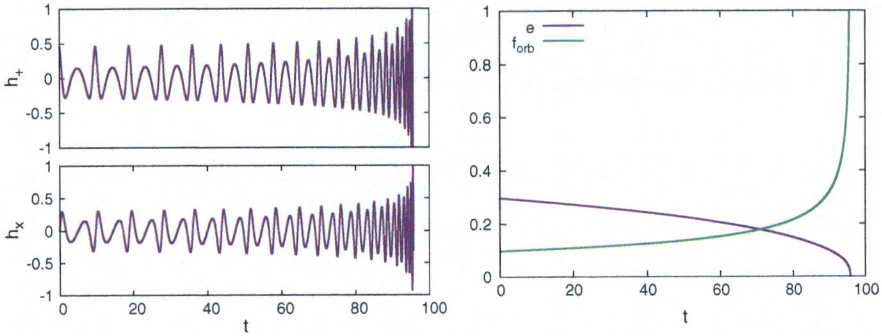

Fig. 35 The panel on the left shows the leading order plus and cross waveforms for an eccentric binary. The panel on the right shows the evolution of the orbital frequency and eccentricity for the same system

however some exotic formation scenarios involving three body effects or gas discs that can result in systems entering the sensitive band of a gravitational wave detector with significant eccentricity, so it is important to include eccentricity in the waveform models so as to be able to detect system from these alternative formation channels. Figure 35 shows the plus and cross waveforms when radiation reaction is included. The evolution of the eccentricity and orbital frequency is also shown.

Going beyond the leading order Newtonian description of the orbital motion introduces qualitatively new effects, the most important being periapse precession, which enters at 1-PN order.

6.5 *Spinning Binaries*

The spin of the component masses can have a significant impact on the gravitational wave signal. Spin-orbits effects first enter at 1.5 PN order, and spin-spin effects enter at 2-PN order. Spin impacts the waveforms in two distinct ways. First, the spins impact the orbital dynamics and gravitational wave emission, and hence the phasing of the waves. The phasing is modified regardless of the relative orientation of the spins and the orbital angular momentum. Second, if the spins and the orbital angular momentum are not aligned (or anti-aligned), the spins and the orbital angular momentum will precess according to Eqs. (148). The precession of the orbital plane causes the inclination angle and polarization angle to vary with time according to Eq. (158), resulting in a modulation of the amplitude of the signal. For spin-precessing systems the phase modulations can oscillate, leading to oscillations in the frequency. Thus information about the spins is encoded in the amplitude modulation (AM) and frequency modulation (FM) of the signal. Note that these modulations lead to a break down of the stationary phase approximation, and more sophisticated methods have to be used to compute the Fourier transforms of the signals.

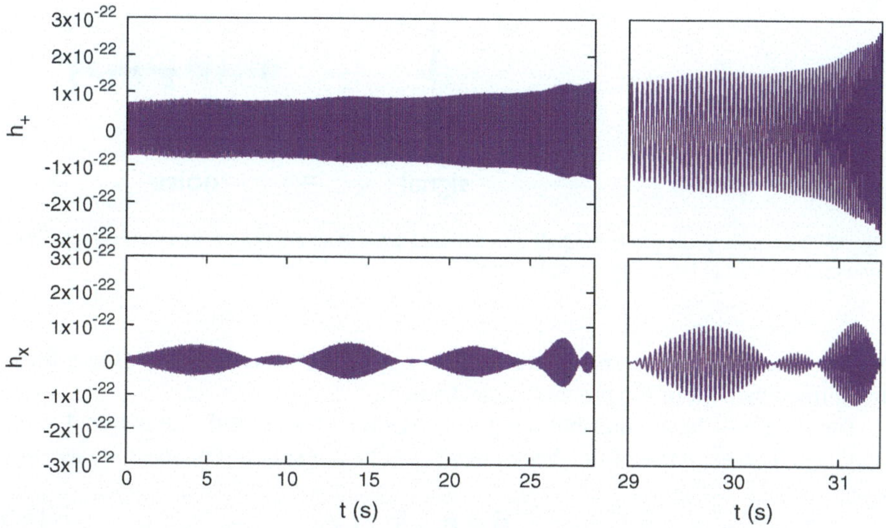

Fig. 36 Plus and cross waveforms at 2.5 PN order for a quasi-circular binary showing the effects of spin precession. The second set of panels zooms in on the signal close to merger. In this example the amplitude modulation is especially pronounced in the cross polarization

Figure 36 shows the plus and cross waveforms for a spinning binary computed at 2.5-PN order. The system shown had $m_1 = 20M_\odot$, $m_2 = 15M_\odot$, $e = 0$ and spin magnitudes $\chi_1 = |\mathbf{S}_1|/m_1^2 = 0.7$ and $\chi_2 = |\mathbf{S}_2|/m_2^2 = 0.5$. The spins were misaligned with the orbital angular momentum such that configuration at $t = 0$ had $\mathrm{acos}(\hat{L} \cdot \hat{S}_1) = 85°$, $\mathrm{acos}(\hat{L} \cdot \hat{S}_2) = 82°$ and $\mathrm{acos}(\hat{S}_1 \cdot \hat{S}_2) = 110°$.

7 Science Data Analysis

The output from a gravitational wave detector is a time series $d(t)$. Usually we have multiple detectors and hence multiple times series. The data can be aggregated into a vector $\mathbf{d}$, with components that are labeled by detector name and a time stamp. Fundamentally, gravitational wave data analysis is time-series analysis, and many of the standard tools of time series analysis get applied, such as band-pass filters, windows, FFTs, spectral estimators, wavelet transforms *etc*. In these lectures I will gloss over these low level (yet essential) data processing steps and focus on the higher level aspects of the analysis.

The literature on gravitational wave data analysis can be befuddling. There is talk of "matched filtering", detection statistics, false alarm rates, time-slides and parameter estimation. I will get to all of those topics, but I will start with a much simpler description in terms of Bayesian inference [23, 51], where everything we need to know is summarized in a single function—the posterior distribution. Bayesian inference requires just two ingredients, the likelihood function—which turns out to

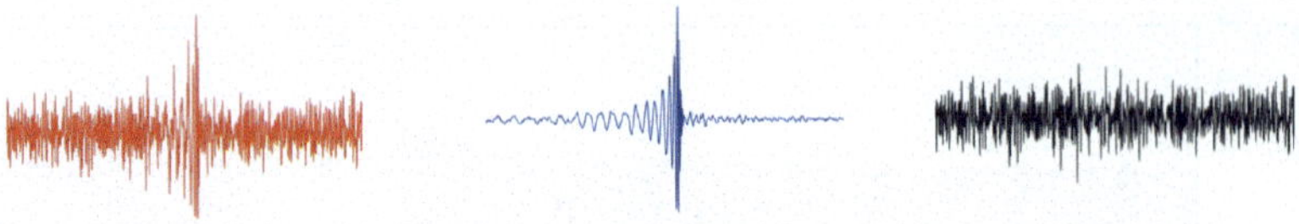

Fig. 37 The basic principle behind all gravitational wave data analysis is that the residual (data minus signal) should be consistent with noise

be the noise model, and a prior, which turns out to be the signal model. Once those are defined the rest of the process is mechanical.

The data will have contributions from detector noise **n** and gravitational wave signals $\mathfrak{h}$. The response of the detectors we will be considering is linear, so we may write

$$\mathbf{d} = \mathbf{n} + \mathfrak{h} \,. \tag{177}$$

Given **d** we would like to infer $\mathfrak{h}$. To do so we need models for the instrument noise and for the gravitational wave signals. For now I will assume that the instrument noise is stochastic, and can be described by some probability distribution $p(\mathbf{n})$. The gravitational wave signal model **h** may be deterministic or stochastic, and might be based on solutions to the Einstein field equations or some more generic model, such as a collection of wavelets. In all cases we demand that the residual $\mathbf{r} = \mathbf{d} - \mathbf{h}$, given by the data minus the model, must be consistent with noise: $\mathbf{r} \sim p(\mathbf{n})$. In other words, *the noise model defines the likelihood function*. This basic principle is illustrated in Fig. 37.

To give a concrete example, consider a Gaussian noise model. The likelihood function is then

$$p(\mathbf{d}|\mathbf{h}, \mathcal{H}) = \frac{1}{(2\pi\, \det\mathbf{C})^{1/2}}\, e^{-\frac{1}{2}(\mathbf{d}-\mathbf{h})\cdot\mathbf{C}^{-1}\cdot(\mathbf{d}-\mathbf{h})}\,, \tag{178}$$

where **C** is the noise correlation matrix. (Recall the way to read notation such as $p(\mathbf{d}|\mathbf{h}, \mathcal{H})$ is: "the probability of observing the data **d** given the presence of a gravitational wave signal **h** under model $\mathcal{H}$"). The exponent in the likelihood is proportional to the chi-squared of the model, and is given by a double sum over detectors I and data samples k:

$$\chi^2(\mathbf{h}) = (\mathbf{d}-\mathbf{h})\cdot\mathbf{C}^{-1}\cdot(\mathbf{d}-\mathbf{h}) = (d_{Ik} - h_{Ik})C^{-1}_{(Ik)(Jm)}(d_{Jm} - h_{Jm})\,. \tag{179}$$

In most cases the noise is uncorrelated between detectors: $C_{(Ik)(Jm)} = \delta_{IJ} S_{Ikm}$, with S_{Ikm} the noise correlation matrix for detector I. An important point is that the noise matrix is itself an a priori unknown quantity that has to be inferred from the data. If the noise is stationary the correlations only depend on the time lag between samples, and the correlation matrix can be diagonalized by transforming to the frequency domain where

$$S_{Ikm} = \delta_{km}\, S_I(f_k)\,. \tag{180}$$

The noise modeling is then reduced to inferring the power spectrum $S_I(f_k)$ for each detector. In some applications, such as pulsar timing, where the data is unevenly sampled in time and the noise is highly non-stationary, the likelihood has to be computed directly in the time domain, and special techniques have to be used to tame the computational cost associated with inverting the large noise correlation matrices.

To find the posterior distribution for the signal model $p(\mathbf{h}, \mathcal{H}|\mathbf{d})$ we apply Bayes theorem:

$$p(\mathbf{h}, \mathcal{H}|\mathbf{d}) = \frac{p(\mathbf{d}|\mathbf{h}, \mathcal{H})p(\mathbf{h}, \mathcal{H})}{p(\mathbf{d}, \mathcal{H})}\,, \tag{181}$$

where $p(\mathbf{h}, \mathcal{H})$ is the prior the defines our signal model and $p(\mathbf{d}, \mathcal{H})$ is the normalization factor

$$p(\mathbf{d}, \mathcal{H}) = \int p(\mathbf{d}|\mathbf{h}, \mathcal{H})p(\mathbf{h}, \mathcal{H})d\mathbf{h}\,. \tag{182}$$

The normalization factor is variously known as the marginal likelihood or the model evidence. Figure 38 shows examples of waveform posteriors computed by the *BayesWave* [14] algorithm for a collection of LIGO/Virgo events.

In many instances we are less interested in the waveforms themselves and more interested in the parameters that define the signals. In Bayesian inference we marginalize over (integrate out) quantities we are not interested in. For example, suppose that we have a model for the signals $p(\mathbf{h}, \boldsymbol{\theta}, \mathcal{H})$ that is described by parameters $\boldsymbol{\theta}$. We can marginalized over $\mathbf{h}$ to arrive at a new (marginal) likelihood that only involves the model parameters:

$$p(\mathbf{d}|\boldsymbol{\theta}, \mathcal{H}) = \int p(\mathbf{d}|\mathbf{h}, \boldsymbol{\theta}, \mathcal{H})p(\mathbf{h}, \boldsymbol{\theta}, \mathcal{H})d\mathbf{h}\,. \tag{183}$$

The posterior distribution for the model parameters follows from Bayes' theorem:

$$p(\boldsymbol{\theta}|\mathbf{d}, \mathcal{H}) = \frac{p(\mathbf{d}|\mathcal{H}, \boldsymbol{\theta})p(\boldsymbol{\theta}|\mathcal{H})}{p(\mathbf{d}|\mathcal{H})}\,, \tag{184}$$

where

$$p(\mathbf{d}|\mathcal{H}) = \int p(\mathbf{d}|\mathcal{H}, \boldsymbol{\theta})p(\boldsymbol{\theta}|\mathcal{H})d\boldsymbol{\theta}\,, \tag{185}$$

is the model evidence and $p(\boldsymbol{\theta}|\mathcal{H})$ is the prior on the signal parameters. Examples of signal models that are used in gravitational wave analyses include theoretical waveform templates for binary mergers, wavelet based models for generic bursts and probabilistic models for stochastic signals. Both the template based models and the

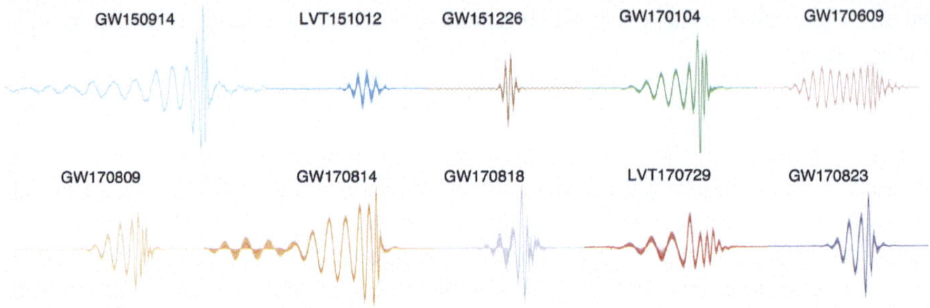

Fig. 38 Waveform posteriors $p(\mathbf{h}, \mathcal{H}|\mathbf{d})$ showing the 90% credible regions of the waveform reconstructions by the *BayesWave* algorithm for a collection of LIGO/Virgo detections

wavelet based models map the gravitational wave signal to a parameterized function $\mathbf{h}(\theta)$:

$$p(\mathbf{h}|\mathcal{H}, \boldsymbol{\theta}) = \delta(\mathbf{h} - \mathbf{h}(\boldsymbol{\theta})) . \tag{186}$$

For Gaussian detector noise the marginal likelihood is then given by

$$p(\mathbf{d}|\boldsymbol{\theta}, \mathcal{H}) = \frac{1}{(2\pi \det \mathbf{C})^{1/2}} e^{-\frac{1}{2}(\mathbf{d}-\mathbf{h}(\boldsymbol{\theta}))\cdot \mathbf{C}^{-1}\cdot(\mathbf{d}-\mathbf{h}(\boldsymbol{\theta}))} . \tag{187}$$

If the noise is stationary and uncorrelated between detectors, the χ^2 term in the likelihood can be written as

$$\chi^2(\boldsymbol{\theta}) = (\mathbf{d} - \mathbf{h}(\boldsymbol{\theta})|(\mathbf{d} - \mathbf{h}(\boldsymbol{\theta})) \tag{188}$$

where we have introduced the noise weighted inner product

$$(\mathbf{a}|\mathbf{b}) = \sum_I \int_0^\infty \frac{2(\tilde{a}_I(f)\tilde{b}_I^*(f) + \tilde{a}_I^*(f)\tilde{b}_I(f))}{S_I(f)} df . \tag{189}$$

Here the inner product has been written in its conventional form in terms of a integral over frequency. In practice the data will always have a finite duration $T_{\rm obs}$ and a finite sample rate dt, and the integral gets replaced by a sum over frequencies $f_k = k/T_{\rm obs}$ from $k = 0$ to $k = T_{\rm obs}/(2dt)$.

For binary mergers the waveform templates $\mathbf{h}(\boldsymbol{\theta})$ are built from the $h_+, h_\times$ polarizations states computed using the techniques discussed in Sect. 6. The source frame waveforms have to be convolved with the instrument response as described in Sect. 4. For a fully general binary black hole system the parameter vector $\boldsymbol{\theta}$ will have 17 components. Seven of the parameters are time invariant: the two masses m_1, m_2; the dimensionless spin magnitudes $\chi_1 = |\mathbf{S}_1|/m_1^2$, $\chi_2 = |\mathbf{S}_2|/m_2^2$, the sky location (θ, ϕ) and the luminosity distance to the source D_L. The other ten parameters have to be referenced to some particular orbital separation, which defines a reference time

t_*. The parameters defined at t_* are the four spin components $\hat{S}_1, \hat{S}_2$; the overall phase ϕ_*; the eccentricity e and periapse angle ϕ_e; and two angles that define the orientation of the total angular momentum vector $\mathbf{J} = \mathbf{L} + \mathbf{S}_1 + \mathbf{S}_2$. There are many alternative ways to parameterize the signals. For example, the spin/orbit parameters can also be described in terms of the angles (θ_L, ϕ_L), (θ_1, ϕ_1) and (θ_2, ϕ_2) between $\mathbf{L}, \mathbf{S}_1, \mathbf{S}_2$ and $\mathbf{J}$ at the reference time. The merger time t_c is often used to set the time reference, and the chirp mass $\mathcal{M}$ and total mass M are often used in place of the individual masses m_1, m_2. Priors can be placed on these parameters using information from past astronomical observations and from theoretical considerations. For example, we might assume that binary systems follow the distribution of galaxies on the sky, which at large distances goes over to a uniform distribution. The range of spin magnitudes can be limited to the region [0, 1] so as to avoid naked singularities.

With wavelet based models the templates are given by a sum of wavelets. For example, the original version of the *BayesWave* algorithm wrote the two polarization states as a sum of Morlet–Gabor continuous wavelets:

$$
\begin{aligned}
h_+(\boldsymbol{\theta})(t) &= \sum_{i=1}^{N} A_i e^{-(t-t_i)^2/\tau_i^2} \cos(2\pi f_i (t - t_i) + \phi_i) \\
h_\times(\boldsymbol{\theta})(t) &= \varepsilon h_+(\boldsymbol{\theta})(t) ,
\end{aligned} \tag{190}
$$

where ε sets the ellipticity of the signal, the extremes being $\varepsilon = \pm 1$ for circular polarization and $\varepsilon = 0$ for linear polarization. The number of wavelets, N, can be varied. The full template $\mathbf{h}(\boldsymbol{\theta})$ folds in the detector response, which brings in a dependance on the sky location and polarization angle. The full parameter vector $\boldsymbol{\theta}$ has dimension $4 + 5N$.

For a stochastic gravitational wave signal, such as might be produced by inflation or other violent processes in the early Universe, the gravitational wave amplitudes $h_+, h_\times$ are random variables. For a Gaussian stochastic signal we are interested in inferring the spectrum $S_h(f)$. The prior on the signal model is then

$$
p(\mathbf{h}|\mathcal{H}, \mathbf{S}_h) = \frac{1}{\sqrt{\det(2\pi \mathbf{S}_h)}} e^{-\frac{1}{2}\mathbf{h}^\dagger \mathbf{S}_h^{-1} \mathbf{h}} . \tag{191}
$$

Theoretical models can provide priors on the shape and amplitude of the spectrum $p(\mathbf{S}_h)$, and on the degree of anisotropy and polarization. For an un-polarized, statistically isotropic, stationary Gaussian stochastic background and un-correlated stationary Gaussian instrument noise, the integral in (183) yields the likelihood [15]

$$
p(\mathbf{d}|\mathbf{S}_h) = \frac{1}{(2\pi \det\mathbf{G})^{1/2}} e^{-\frac{1}{2}(\mathbf{d}\cdot\mathbf{G}^{-1}\cdot\mathbf{d})} , \tag{192}
$$

where

$$
G_{(Ik),(Jm)} = \delta_{km} \left(\delta_{IJ}\, S_I(f_k) + \gamma_{IJ}(f_k)\, S_h(f_k)\right) . \tag{193}
$$

Here $\gamma_{IJ}(f_k)$ describes how the common gravitational wave signal is correlated between detectors [49]. In the interferometry literature it is called the overlap reduction function, while in the pulsar timing literature it is called the Hellings–Downs curve. The stochastic signal can be separated from the stochastic noise since the signal is correlated between detectors, while the noise is not.

7.1 Posterior Distributions, Bayesian Learning and Model Evidence

The posterior distribution $p(\boldsymbol{\theta}|\mathbf{d},\mathcal{H})$ and the model evidence $p(\mathbf{d}|\mathcal{H})$ summarize everything about the model $\mathcal{H}$ that can be gleaned from the data. The posterior distribution can be used to compute point estimates for each parameter such as the mean, median and mode, as well as credible intervals for one or more parameters. For example, the mean value for parameter θ^j is given by

$$\bar{\theta}^j = \int \theta^j \, p(\boldsymbol{\theta}|\mathbf{d},\mathcal{H})\, d\boldsymbol{\theta} \,. \tag{194}$$

The mode, or peak, of the posterior distribution defines the *maximum a posteriori* (MAP) parameter values $\boldsymbol{\theta}_{\rm MAP}$. In many instances the posterior distribution will be multi-modal, with multiple local maxima.

In typical gravitational wave analyses the parameter dimension is large, and it is not possible to plot the full posterior distribution. Instead lower dimensional marginal distributions are shown such as the probability distribution for a single parameter θ^j:

$$p(\theta^j) = \int p(\boldsymbol{\theta}|\mathbf{d},\mathcal{H}) \prod_{k\neq j} d\theta^k \,, \tag{195}$$

or the joint probability distribution for a pair of parameters θ^i, θ^j:

$$p(\theta^i,\theta^j) = \int p(\boldsymbol{\theta}|\mathbf{d},\mathcal{H}) \prod_{k\neq i,j} d\theta^k \,. \tag{196}$$

It has become common practice to display posterior distributions using "corner plots" that show the joint distribution for each pair of parameters along with the marginal distribution for each individual parameter. An example of such a corner plot is shown in Fig. 39 derived from a simulation of a spinning binary black hole inspiral observed by the advanced LIGO/Virgo instruments. It is difficult to fit all $d(d+1)/2 = 120$ corner plot panels for a $d = 15$ dimensional posterior into a single graph, so I have instead selected 11 of the more interesting parameters and shown them in two corner plots with 5 and 6 rows respectively.

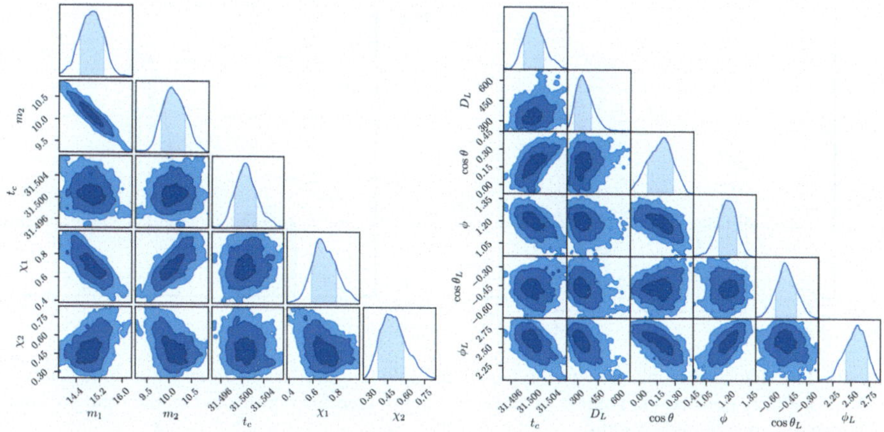

Fig. 39 One and two dimensional marginalize posterior distributions for a simulated spinning binary black hole inspiral observed by the advanced LIGO/Virgo instruments. The colored regions in the two dimensional plots show the "Gaussian equivalent" 1-sigma, 2-sigma and 3-sigma regions. The shaded region in the one dimensional plots shows the 68% "1-sigma" credible regions for each parameter

In Fig. 39 the posterior density is colored in terms of credible regions that contain a designated fraction of the total probability. There are many different ways to define credible regions. Two of the more popular choices at central credible regions and minimum volume central regions. The credible regions show in Fig. 39 are of the minimum volume variety. Figure 40 illustrates the central and minimum area credible regions for a one dimensional probability distribution. The boundaries of the central credible region, $x_{\rm min}$ and $x_{\rm max}$, that contain a fraction α of the total probability are defined:

$$\int^{x_{\rm min}} p(x)dx = \int_{x_{\rm max}} p(x)dx = \frac{\alpha - 1}{2}\,. \tag{197}$$

Central credible intervals are often used to quote the uncertainties on individual parameters. For example, the mass parameters for the example shown in Fig. 39 can be quoted in terms of the mean values and the boundaries of the 1-sigma equivalent credible region: $m_1 = (14.95^{+0.38}_{-0.44})M_\odot, m_2 = (10.04^{+0.31}_{-0.21})M_\odot$. Note that the credible intervals in this case are not symmetric about the mean values.

The minimum volume credible region can be made up of multiple disjoint parts. It is defined by a minimum density p_* such that

$$\int [p(x) \geq p_*]dx = \alpha\,. \tag{198}$$

In other words, only regions with densities above p_* are included in the minimum volume credible region. By construction, the volume (here the length of the region

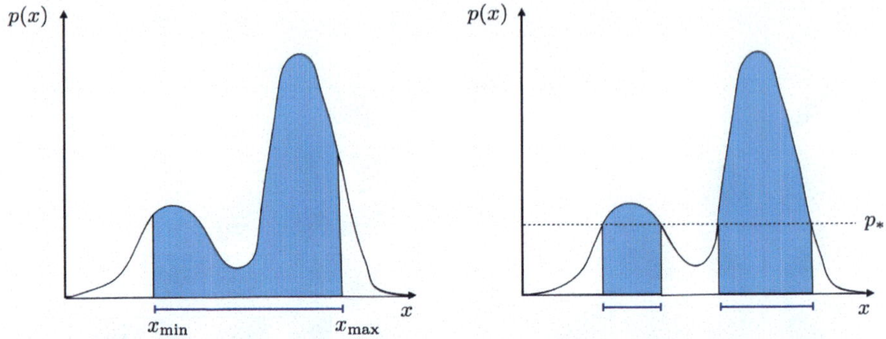

Fig. 40 The shaded regions show two types of credible regions. Each contains the same fraction of the probability density. On the left we have the central credible region, and on the right we have the minimum area credible region

with $p(x) \geq p_*$) will always less than or equal to that of any other credible region containing the same total probability.

Bayesian inference is a pure expression of the scientific method. Bayes' theorem describes how we learn from data, and the Bayesian odds ratio quantifies our belief in competing hypotheses. Bayes theorem describes how our prior belief $p(\boldsymbol{\theta})$ is updated to our posterior belief $p(\boldsymbol{\theta}, \mathbf{d}_1)$ after incorporating, via the likelihood function, the information contained in data $\mathbf{d}_1$. The updated probability distribution for $\boldsymbol{\theta}$ becomes our new prior, which can then be updated by additional data $\mathbf{d}_2$ to give the posterior distribution $p(\boldsymbol{\theta}, \mathbf{d}_1, \mathbf{d}_2)$. Note that we get the same result if we start with data $\mathbf{d}_2$ to arrive at the posterior $p(\boldsymbol{\theta}, \mathbf{d}_2)$, which then serves as the prior when incorporating data $\mathbf{d}_1$ to yield $p(\boldsymbol{\theta}, \mathbf{d}_2, \mathbf{d}_1) = p(\boldsymbol{\theta}, \mathbf{d}_1, \mathbf{d}_2)$. This is true whether the data are independent or dependent. The amount we learn about a hypothesis $\mathcal{H}$ from data $\mathbf{d}$ can be quantified in terms of how different the posterior distribution is from the prior distribution, which we can measured using the Kullback–Leibler divergence:

$$I_{\rm KL} = \int p(\boldsymbol{\theta}|\mathbf{d}, \mathcal{H}) \log_2 \left(\frac{p(\boldsymbol{\theta}|\mathbf{d}, \mathcal{H})}{p(\boldsymbol{\theta}|\mathcal{H})} \right) d\boldsymbol{\theta} \quad \text{(bits)}. \tag{199}$$

If the posterior distribution is much more concentrated than the prior distribution then the information gain is large. Conversely, if someone has very strong prior beliefs, they learn little, even when confronted with a vast amount of evidence.

The probability that a model $\mathcal{H}_k$ describes the data $\mathbf{d}$ is given by Bayes' theorem:

$$p(\mathcal{H}_k|\mathbf{d}) = \frac{p(\mathbf{d}|\mathcal{H}_k) p(\mathcal{H}_k)}{p(\mathbf{d}) = \sum_j p(\mathbf{d}|\mathcal{H}_j) p(\mathcal{H}_j)} . \tag{200}$$

Here $p(\mathbf{d}|\mathcal{H}_k)$ is the marginal likelihood, or evidence for model $\mathcal{H}_k$ and $p(\mathcal{H}_k)$ is our prior belief in the model. The numerator is a normalizing factor that sums (or integrates) over all possible models. In most applications it is impossible to write down all possible models, and we instead consider the Bayesian odds ratios between pairs of competing models:

$$\mathcal{O}_{ij} = \frac{p(\mathcal{H}_i|\mathbf{d})}{p(\mathcal{H}_j|\mathbf{d})} = \left(\frac{p(\mathbf{d}|\mathcal{H}_i)}{p(\mathbf{d}|\mathcal{H}_j)}\right)\left(\frac{p(\mathcal{H}_i)}{p(\mathcal{H}_j)}\right) = \mathcal{B}_{ij}\mathcal{P}_{ij}\,. \tag{201}$$

The unknown normalization factor cancels out in the odds ratio. The first term in parentheses is the evidence ratio, or Bayes factor $\mathcal{B}_{ij}$ for the two models, while the second term in parentheses is the prior odds ratio $\mathcal{P}_{ij}$. The Bayes factor can be used to measure the significance of an event by comparing the evidence for the noise-only model to the evidence for the signal+noise model. A word of caution however, is that utility of such Bayes factors are only as good as the models. For example, if you are using a Gaussian model for the noise, when in reality the noise is non-Gaussian, then the Bayes factors between the noise-only and signal+noise model will not be a reliable measure of whether an astrophysical signal is indeed present in the data.

7.2 *Maximum Likelihood and the Fisher Information Matrix*

Before continuing the discussion of signal detection and parameter estimation, it is instructive to digress a little and consider maximum likelihood parameter estimation and error forecasting using the Fisher Information Matrix. These topics are usually discussed in the classical, or frequentist, approach to gravitational wave detection, but they are also closely related to Taylor series expansion of Bayesian posterior distributions.

To simplify the discussion we will assume the noise is stationary and gaussian with a known spectrum. The log likelihood is then proportional to the chi-squared given in Eq. (188), and local maxima of the likelihood can be found by setting the derivative with respect to the model parameters to zero. The maximum likelihood solution can be found by Taylor expanding the signal model about the true parameters $\boldsymbol{\theta}_T$.

$$\mathbf{h}(\boldsymbol{\theta}) = \mathbf{h}_T + \partial_i \mathbf{h}_T \Delta\theta^i + \frac{1}{2}\partial_i\partial_j \mathbf{h}_T \Delta\theta^i \Delta\theta^j + \mathcal{O}\left(\Delta\theta^3\right). \tag{202}$$

Here we are using the shorthand notation $\mathbf{h}_T = \mathbf{h}(\boldsymbol{\theta}_T) = \mathfrak{h}$ and $\partial_i \mathbf{h}_T = \partial_{\theta^i}\mathbf{h}(\boldsymbol{\theta})|_{\theta=\theta_T}$. The Taylor expansion of the chi-squared is then

$$\chi^2(\boldsymbol{\theta}) = (\mathbf{n}|\mathbf{n}) - 2(\mathbf{n}|\partial_i \mathbf{h}_T)\Delta\theta^i + \left[(\partial_i \mathbf{h}_T|\partial_j \mathbf{h}_T) - (\mathbf{n}|\partial_i\partial_j \mathbf{h}_T)\right]\Delta\theta^i\Delta\theta^j + \mathcal{O}(\Delta\theta^3)\,. \tag{203}$$

Setting $\partial_k \chi^2(\boldsymbol{\theta}) = 0$ we find

$$\Delta\theta^k_{\rm ML} = \Gamma^{kl}(\mathbf{n}|\partial_l \mathbf{h}_T)\,, \tag{204}$$

where $\Gamma_{kj} = (\partial_k \mathbf{h}_T | \partial_l \mathbf{h}_T)$ is the Fisher information matrix, and $\Gamma^{kj} = \Gamma^{-1}_{kj}$ is its matrix inverse. In arriving at Eq. (204) we have dropped terms coming from the inner product $(\mathbf{n}|\partial_k \partial_l \mathbf{h}_T)$ since they are of order $\Delta\theta$, and keeping them generates higher order corrections. What Eq. (204) tells us is that best-fit parameter values are perturbed away from the true parameter values by an amount that depends on how well variations in the signal can mimic noise. Put another way, the maximum likelihood solution is able to achieve a higher likelihood by fitting some of the noise with the signal model.

In the frequentist approach one considers multiple repetitions of the measurement and computes expectation values. Assuming zero mean, gaussian noise we have $\mathrm{E}[\tilde{n}(f)] = 0$ and $\mathrm{E}[\tilde{n}(f)\tilde{n}^*(f')] = \frac{1}{2} S_n(f)\delta(f - f')$. Using these expressions is it easy to show that

$$\mathrm{E}[\Delta\theta^k_{\mathrm{ML}}] = 0\,, \tag{205}$$

and

$$\mathrm{E}[\Delta\theta^k_{\mathrm{ML}} \Delta\theta^l_{\mathrm{ML}}] = \Gamma^{kl}\,. \tag{206}$$

In other words, the parameter errors should follow a Gaussian distribution with zero mean, and with a covariance matrix given by the inverse of the Fisher information matrix.

The frequentist viewpoint is reasonable if you are interested in forecasting the performance of an observatory, but it makes no sense to consider multiple realizations of an actual observation: you can not expect the Universe to repeat the same black hole merger multiple times. For actual observations it makes more sense to take a Bayesian approach. The best-fit parameters (in a maximum likelihood sense) are obtained in the same way, but now we interpret Eq. (204) to be the perturbation due to the actual noise realization present in the data. We can expand about the maximum likelihood solution by writing $\Delta\theta^k = \Delta\theta^k_{\mathrm{ML}} + \delta\theta^k$ and re-expanding the chi-squared:

$$\chi^2(\boldsymbol{\theta}) = \chi^2_{\mathrm{ML}} + \Gamma_{ij}\delta\theta^i\delta\theta^j + \mathcal{O}(\delta\theta^3)\,. \tag{207}$$

where $\chi^2_{\mathrm{ML}} = (\mathbf{n}|\mathbf{n}) - \Gamma^{ij}(\mathbf{n}|\partial_i \mathbf{h}_T)(\mathbf{n}|\partial_j \mathbf{h}_T)$ is the maximum likelihood value for the chi-squared. Rather than evaluate the derivatives in the Fisher matrix at the unknown true parameter values, we can evaluate them at the known maximum likelihood parameter values since the difference between the two is next order in the expansion. Thus we can approximate the likelihood in the vicinity of the maximum likelihood solution as

$$p(\mathbf{d}|\mathcal{H}, \boldsymbol{\theta}) \simeq \sqrt{\det(\boldsymbol{\Gamma}/2\pi)}\, e^{-\frac{1}{2}\Gamma_{ij}\delta\theta^i\delta\theta^j}\,. \tag{208}$$

We can perform a similar expansion of the posterior distribution about the MAP parameter values:

$$p(\boldsymbol{\theta}|\mathbf{d}, \mathcal{H}) \simeq \sqrt{\det(\boldsymbol{\Upsilon}/2\pi)}\, e^{-\frac{1}{2}\Upsilon_{ij}\delta\theta^i\delta\theta^j}\,. \tag{209}$$

where

$$\Upsilon_{ij} = \Gamma_{ij} - \partial_i \partial_j \ln p(\boldsymbol{\theta}|\mathcal{H}) \,. \tag{210}$$

For flat priors the second term vanishes and the posterior matches the likelihood. The quadratic expansion of the posterior (209) is a multi-variate Gaussian distribution with correlation matrix $\boldsymbol{\Upsilon}^{-1}$. The end result is that the Bayesian and frequentist approaches yield similar results in this instance, even though the philosophy behind the two approaches is very different.

The expansion about the MAP parameters can also be used to estimate the model evidence using the Laplace approximation for exponential integrals:

$$p(\mathbf{d}|\mathcal{H}) \approx p(\mathbf{d}|\mathcal{H}, \boldsymbol{\theta}_{\rm MAP}) \frac{p(\boldsymbol{\theta}_{\rm MAP}|\mathcal{H})}{\sqrt{\det(\boldsymbol{\Upsilon}/2\pi)}} \,. \tag{211}$$

The term $1/\sqrt{\det(\boldsymbol{\Upsilon}/2\pi)}$ can be interpreted as the one-sigma posterior volume, V_σ. For uniform priors, $p(\boldsymbol{\theta}|\mathcal{H}) = V^{-1}$, where V is the prior volume and $p(\mathbf{d}|\mathcal{H}, \boldsymbol{\theta}_{\rm MAP})$ is equal to the maximum likelihood $\mathcal{L}_{\rm max}$. Thus for uniform (or slowly varying) priors, we have

$$p(\mathbf{d}|\mathcal{H}) \approx \mathcal{L}_{\rm max} \left(\frac{V_\sigma}{V} \right) \,. \tag{212}$$

What this tell us is that the evidence is higher for models that fit the data better (higher $\mathcal{L}_{\rm max}$), but there is a Occam penalty (the second term in the above equation) that works against models that have a large number of parameters.

7.3 Frequentist Detection Statistics

Suppose that we have two models, one where a signal is present, $\mathcal{H}_1$, and one where there is only noise $\mathcal{H}_0$. Assuming uniform priors and using the Laplace approximation, the log Bayes factor between the model is given by

$$\begin{aligned} \log \mathcal{B}_{10} = \frac{p(\mathbf{d}|\mathcal{H}_1)}{p(\mathbf{d}|\mathcal{H}_0)} &\approx \log \left(\frac{p(\mathbf{d}|\mathcal{H}_1, \boldsymbol{\theta}_{\rm ML})}{p(\mathbf{d}|\mathcal{H}_0)} \right) + \text{Occam terms} \\ &\approx \log \mathcal{L}_{\rm max} + \text{Occam terms} \end{aligned} \tag{213}$$

where $\log \mathcal{L}_{\rm max}$ is the maximum of the log likelihood ratio. For Gaussian noise the log likelihood ratio is given by

$$\begin{aligned} \log \mathcal{L} &= \frac{1}{2} \left((\mathbf{d} - \mathbf{h}|\mathbf{d} - \mathbf{h}) - (\mathbf{d}|\mathbf{d}) \right) \\ &= (\mathbf{d}|\mathbf{h}) - \frac{1}{2}(\mathbf{h}|\mathbf{h}) \,. \end{aligned} \tag{214}$$

A gravitational wave template $\mathbf{h}$ can always be written in terms of an overall amplitude ρ such that $\mathbf{h} = \rho\hat{\mathbf{h}}$ where $(\hat{\mathbf{h}}|\hat{\mathbf{h}}) = 1$. If we maximize the log likelihood ratio with respect to ρ by setting $\partial_\rho \log \mathcal{L} = 0$ we find

$$\rho = (\mathbf{d}|\hat{\mathbf{h}}) \,, \tag{215}$$

and

$$\log \mathcal{L} = \frac{1}{2}\rho^2 \,. \tag{216}$$

The quantity ρ is referred to as the matched filter detection statistic, and can be used to define the matched filter signal-to-noise ratio. In the absence of the signal, $\mathfrak{h} = 0$, we have

$$\begin{aligned} \mathrm{E}[\rho_{\mathfrak{h}=0}] &= 0 \\ \mathrm{Var}[\rho_{\mathfrak{h}=0}] &= \left(\mathrm{E}[\rho^2_{\mathfrak{h}=0}] - \mathrm{E}^2[\rho_{\mathfrak{h}=0}]\right) = (\hat{\mathbf{h}}|\hat{\mathbf{h}}) = 1 \,. \end{aligned} \tag{217}$$

When a signal is present and it is matched by the the filter, $\mathbf{h} = \mathfrak{h}$, we have

$$\mathrm{E}[\rho_{\mathbf{h}=\mathfrak{h}}] = (\mathfrak{h}|\mathfrak{h})^{1/2} \tag{218}$$

giving an expected signal-to-noise ratio of

$$\mathrm{SNR} = \frac{\mathrm{E}[\rho_{\mathbf{h}=\mathfrak{h}}]}{\mathrm{Var}[\rho_{\mathfrak{h}=0}]^{1/2}} = (\mathfrak{h}|\mathfrak{h})^{1/2} \,. \tag{219}$$

The maximization of the likelihood can be performed algebraically with respect to an overall constant amplitude ρ and phase ϕ_0 by writing

$$\mathbf{h}(\boldsymbol{\lambda}) = \rho\mathbf{h}_c(\boldsymbol{\kappa})\cos(\phi_0) + \rho\mathbf{h}_s(\boldsymbol{\kappa})\sin(\phi_0) \tag{220}$$

where $(\mathbf{h}_c, \mathbf{h}_c) = (\mathbf{h}_s|\mathbf{h}_s) = 1$, $(\mathbf{h}_c|\mathbf{h}_s) = 0$ and $\boldsymbol{\kappa} = \boldsymbol{\lambda}/\{\rho, \phi_0\}$. The log likelihood ratio then becomes

$$\begin{aligned} \log \mathcal{L}(\boldsymbol{\lambda}) &= \rho(\mathbf{d}|\mathbf{h}_c(\boldsymbol{\kappa}))\cos(\phi_0) + \rho(\mathbf{d}|\mathbf{h}_s(\boldsymbol{\lambda}))\sin(\phi_0) - \frac{1}{2}\rho^2 \\ &= \rho\varrho(\boldsymbol{\kappa})\cos(\phi_0 - \varphi(\boldsymbol{\kappa})) - \frac{1}{2}\rho^2 \,. \end{aligned} \tag{221}$$

where

$$\varrho(\boldsymbol{\kappa}) = \sqrt{(\mathbf{d}|\mathbf{h}_c(\boldsymbol{\kappa}))^2 + (\mathbf{d}|\mathbf{h}_s(\boldsymbol{\kappa}))^2} \,, \tag{222}$$

and

$$\varphi(\boldsymbol{\kappa}) = \arctan\left(\frac{(\mathbf{d}|\mathbf{h}_s(\boldsymbol{\kappa}))}{(\mathbf{d}|\mathbf{h}_c(\boldsymbol{\kappa}))}\right) \,. \tag{223}$$

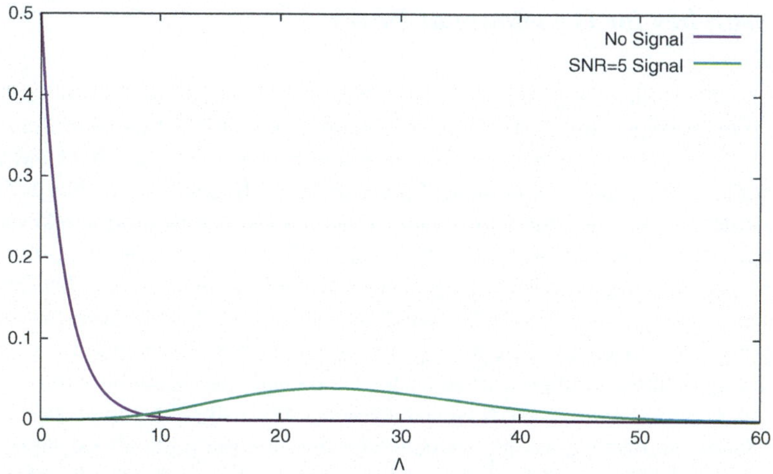

Fig. 41 Probability distributions for the Λ-statistic in Gaussian noise for pure noise, and for data containing a SNR = 5 signal

The likelihood is maximized by setting $\phi_0 = \varphi(\boldsymbol{\kappa})$ and $\rho = \varrho(\boldsymbol{\kappa})$ so that $\log \mathcal{L}_{\max\,\{\rho,\phi_0\}} = \varrho^2(\boldsymbol{\kappa})/2$. The quantity $\Lambda(\boldsymbol{\kappa}) = 2 \log \mathcal{L}_{\max\,\{\rho,\phi_0\}} = \varrho^2(\boldsymbol{\kappa})$ can be shown to follow a non-central chi-squared distribution with two degrees of freedom, $\Lambda \sim \chi_2^2(\mathrm{SNR}^2)$, with non-centrality parameter equal to the signal-to-noise ratio squared $\mathrm{SNR}^2 = (\mathfrak{h}|\mathfrak{h})$. Absent a signal, the distribution reduces to a central chi-squared distribution with two degrees of freedom, otherwise known as a Rayleigh distribution.

Figure 41 shows the theoretical probability distribution for the Λ detection statistic for pure noise, and for data containing a SNR = 5 signal. A *false alarm* occurs when we incorrectly conclude there is a signal present. Setting a false alarm probability of <1% corresponds to requiring that $\Lambda > 13.28$. A *false dismissal* occurs when we conclude there is no signal when one is indeed present. Using a 1% false alarm threshold implies that there is a 4.38% chance that we will falsely dismiss a SNR = 5 signal. For the first detections of gravitational waves the LIGO and Virgo collaborations were very conservative, and demanded that the *false alarm rate*, *i.e.* the number of random events mistaken for gravitational wave signals per unit time, should be very small. Setting a false alarm rate (FAR) of one per 100,000 years over a one year stretch of observation corresponds to a false alarm probability of $p_{\mathrm{FA}} = T_{\mathrm{obs}} \times \mathrm{FAR} = 10^{-5}$.

An important caveat to the preceding discussion is that the distribution shown for the $\Lambda(\boldsymbol{\kappa})$ pertains for *fixed* values of the parameters $\boldsymbol{\kappa}$. In an actual search the correct parameter values are not known a priori, and they must be searched over to find the values that maximize the log likelihood. The probability distribution for the search statistic maximized over all parameters, $\Lambda_{\max}$, does not follow a Rayleigh distribution, and in all but the simplest cases must be computed numerically.

7.4 Searches for Gravitational Waves

The approach to detecting gravitational waves vary between and within collaborations. For example, the Parkes Pulsar Timing Array (PPTA) collaboration have traditionally used frequentist techniques, as have groups within the LIGO and Virgo collaborations that search for compact binary mergers. In contrast, the North American NanoHertz Gravitational Observatory (NANOGrav) collaboration takes a predominantly Bayesian approach, as does the LIGO-Virgo `BayesWave` group. For strong signals the likelihood is highly peaked around the maximum value, and the Bayesian and frequentist approaches yield very similar results. To keep the discussion focused, I will restrict attention to the template based searches for compact binary mergers of the kind performed by the LIGO and Virgo collaborations. As we saw in the previous section, the Bayesian evidence for a signal being present in stationary, Gaussian data can be approximated by the maximum likelihood statistic Λ. In practice the LIGO and Virgo data are not perfectly stationary and Gaussian, and slightly different search statistics are used, and the probability distribution for these statistics under the noise hypothesis are derived empirically from the data. Since it is not known a priori if a given stretch of data contains a signal, the noise properties are determined by first scrambling the data to remove the possibility of detecting a signal. This is done by introducing relative time shifts to the data that are greater than the light travel time between the detectors, ensuring that any signals present appear as noise fluctuations in the shifted data. Just a few weeks of data can be used to simulate millions of years of signal-free observation, making it possible to estimate the probability distribution for the noise down to very small false alarm probabilities (or false alarm rates).

To compute the maximum likelihood statistic Λ (or its equivalent), the likelihood has to be computed and maximized. For some parameters the maximization can be performed algebraically, using the procedure described in Eqs. (220)–(223), while for others a direct search has to be performed. The direct search is usually performed by discretizing the parameter space and performing a grid search with a bank of templates. The spacing of the grid is chosen such that adjacent templates have significant overlap to ensure that signals that lie between grid points are not missed. When the inner-products in the likelihood are computed in the Fourier domain, the Λ statistic can be maximized with respect to the overall time offset t_0 by computing the complex time series

$$z(\boldsymbol{\kappa}', t_0) = 4\int_0^\infty \frac{\tilde{d}(f)\tilde{h}_c(\boldsymbol{\kappa}; f)}{S_n(f)}\, e^{2\pi f t_0} df \;=\; \varrho(\boldsymbol{\kappa}', t_0)e^{i\varphi(\boldsymbol{\kappa}', t_0)} \tag{224}$$

where $\boldsymbol{\kappa}' = \boldsymbol{\kappa}/\{t_0\} = \boldsymbol{\lambda}/\{\rho, \phi_0, t_0\}$. The complex time series can be efficiently computed by an inverse fast Fourier transform (iFFT), and the maximum value of $\varrho(\boldsymbol{\kappa}, t_0)$ read off directly from the iFFT. The maximization procedure can be applied to networks of detectors by extending the inner products to a sum over the detector network, and likewise generalizing Eq. (224). With a network of detectors the maximization over ρ, ϕ_0 can be be generalized to cover a larger set of parameters, including the inclination and polarization angles, using an extension of the method described here, yielding a quantity known as the F-statistic. The overall time shift t_0 can still be maximized over, but the difference in arrival times between detector form part of the collection of parameters $\boldsymbol{\kappa}'$ that still have to be searched over. In principle the method described here can be used to maximize t_0 over the entire observation time, but in practice the data is broken up into smaller chunks and the maximization performed on each chunk separately. This is done because the data has occasional gaps when one or more instruments are off-line, and because the noise is not perfectly stationary, so the spectral densities $S_n(f)$ change over time and have to be computed for each chunk. Another benefit of performing the search on short stretches of data is that signals can be picked up in near real time, allowing alerts to be sent out so that electromagnetic observatories can look for counterparts to the gravitational wave events.

Once the likelihood has been maximized with respect to amplitude, phase and arrival time, it remains to maximize over parameters that control the shape of the signal, such as the masses and spins of a binary system. This is typically done using a grid search over a bank of unit normalized waveform templates. Ideally the grid is laid out to provide uniform coverage of the signal space, with a spacing that ensures that only a very small fraction of potentially detectable signals are missed. In practice it is difficult to achieve perfectly uniform spacing, and the computational cost of evaluating the likelihood at each point on the grid requires a trade off between coverage and speed. A useful analogy is that of a trawler pulling a fishing net. If the holes in the net are too large then smaller fish will not get caught, but if the holes in the net are too small the drag will slow down the trawler and limit the number of fish that can be caught. The placement of the grid is guided by considering the overlap between two signals with parameters $\boldsymbol{\lambda}$ and $\boldsymbol{\theta}$, expressed in terms of the *match*

$$\mathrm{M}(\boldsymbol{\lambda}, \boldsymbol{\theta}) = \frac{(\mathbf{h}(\boldsymbol{\lambda})|\mathbf{h}(\boldsymbol{\theta}))}{\sqrt{(\mathbf{h}(\boldsymbol{\lambda})|\mathbf{h}(\boldsymbol{\lambda})(\mathbf{h}(\boldsymbol{\theta})|\mathbf{h}(\boldsymbol{\theta})}} . \tag{225}$$

For nearby signals we can write $\boldsymbol{\theta} = \boldsymbol{\lambda} + \Delta\boldsymbol{\lambda}$ and Taylor expand in $\Delta\lambda^\mu$:

$$\mathrm{M}(\boldsymbol{\lambda}, \boldsymbol{\lambda} + \Delta\boldsymbol{\lambda}) = 1 - \frac{1}{2}\left(\frac{(h_{,\mu}|h_{,\nu})}{(h|h)} - \frac{(h|h_{,\mu})(h|h_{,\nu})}{(h|h)^2}\right)\Delta\lambda^\mu\Delta\lambda^\nu + \cdots . \tag{226}$$

The quantity in brackets is called the template metric $g_{\mu\nu}$, which defines a distance measure in the Riemannian geometry associated with the inner product $(\mathbf{a}|\mathbf{b})$ [41, 42]. We recognize the first term in the template metric as the Fisher matrix divided by the signal-to-noise ratio squared. Using $\mathbf{h} = \rho\hat{\mathbf{h}}$ we see that the second term is equal to $\Gamma_{\rho\mu}\Gamma_{\rho\nu}/(\rho^2\Gamma_{\rho\rho})$ so that

$$
\begin{aligned}
g_{\mu\nu} &= \frac{(h_{,\mu}|h_{,\nu})}{(h|h)} - \frac{(h|h_{,\mu})(h|h_{,\nu})}{(h|h)^2} \\
&= \frac{1}{\rho^2}\left(\Gamma_{\mu\nu} - \frac{\Gamma_{\rho\mu}\Gamma_{\rho\nu}}{\Gamma_{\rho\rho}}\right) .
\end{aligned} \tag{227}
$$

We recognize the term in brackets in the second line of the above equation to be the Fisher matrix projected onto a sub-space that is independent of the overall amplitude ρ. Since the likelihood is directly maximized with respect to ϕ_0 and t_0 these terms can be removed from the template metric using a sequence of projections:

$$
g'_{\mu\nu} = g_{\mu\nu} - \frac{g_{\phi_0\mu} g_{\phi_0\nu}}{g_{\phi_0\phi_0}} \tag{228}
$$

and

$$
g''_{\mu\nu} = g'_{\mu\nu} - \frac{g'_{t_0\mu} g'_{t_0\nu}}{g'_{t_0\phi_0}} . \tag{229}
$$

The match, maximized over amplitude, phase and time offset, is equal to the *fitting factor*

$$
\mathrm{FF} = 1 - \frac{1}{2} g''_{\mu\nu} \Delta\lambda^\mu \Delta\lambda^\nu , \tag{230}
$$

which defines the fraction of the signal-to-noise ratio of the signal $\mathbf{h}(\boldsymbol{\lambda})$ that can be captured by the template $\mathbf{h}(\boldsymbol{\lambda} + \Delta\boldsymbol{\lambda})$. Since the signal-to-noise ratio scales inversely with distance, and since the volume of space grows as the cube of the distance, the fraction of detectable events captured by the grid search scales as FF^3. Demanding that at least 90% of events are detected sets a threshold of $\mathrm{FF} \sim 0.97$. Placing the templates on a hyper-cubic lattice to ensure a $\geq$FF overlap yields cells with volume [41, 42]

$$
\Delta V = 2^d \left(\frac{(1-\mathrm{FF})}{d}\right)^{d/2} \tag{231}
$$

where $d = \dim(\boldsymbol{\kappa}') = \dim(\boldsymbol{\lambda}) - 3 = D - 3$. The total number of templates required is equal to the total parameter volume $V = \int \sqrt{g''}\, d^d\kappa$ divided by the cell size ΔV.

7.5 Bayesian Parameter Estimation

We have seen that Bayesian inference can be used to compute posterior distributions for the gravitational waveforms $\mathbf{h}(\boldsymbol{\theta})$ and the parameters $\boldsymbol{\theta}$ that describe the signal model, and additionally the model evidence. Bayes' theorem tells us that once the signal model and likelihood are defined and the prior distributions specified, the calculation of the posterior distributions and evidence comes down to computing a challenging multi-dimensional integral. It is only in the last two decades that efficient computational techniques, coupled with a increase in micro-processor speed, have made it possible to carry out the necessary computations for real-world applications. Bayesian inference is now rapidly supplanting classical (frequentist) statistics in many branches of science, including gravitational wave astronomy. There are two main approaches used to carry out the Bayesian computation. The first is the Markov Chain Monte Carlo (MCMC) approach [7, 24], which has its roots in statistical mechanics, and the second is Nested Sampling [52], which uses a stochastic Lebesgue integration technique. The MCMC approach produces samples from the posterior distribution without directly evaluating the evidence integral, while Nested Sampling computes the evidence without directly sampling the posterior distribution. With a little extra work the MCMC approach can be used to compute the evidence, and the posterior distributions can be recovered as a by-product of the Nested Sampling approach, so both methods provide a comprehensive framework in which to carry out Bayesian inference. In my own research I exclusively use the MCMC approach as I find it to be better suited to the kinds of models I work with, which are generally of the trans-dimensional variety. Trans-dimensional modeling expands the usual sampling of model parameters to sampling across models in a large model space. Here I will focus on the MCMC approach as it is the one I am most familiar with.

A Markov process is a stochastic process where the current state depends only on the previous state. A Markov process is uniquely defined by the transition probability $p(\mathbf{x}|\mathbf{y})$ from state $\mathbf{x}$ to state $\mathbf{y}$, and is characterized by a unique stationary distribution $\pi(\mathbf{x})$ if the transitions are reversible and satisfy detailed balance $p(\mathbf{x}, \mathbf{y}) = p(\mathbf{y}|\mathbf{x})\pi(\mathbf{x}) = p(\mathbf{x}|\mathbf{y})\pi(\mathbf{y})$, and are additionally aperiodic and positive recurrent (so that the return time to a given state is finite). The transition probability can be factored into the product of a proposal distribution $q(\mathbf{y}|\mathbf{x})$ and an acceptance probability $H(\mathbf{y}|\mathbf{x})$:

$$p(\mathbf{y}|\mathbf{x}) = q(\mathbf{y}|\mathbf{x})H(\mathbf{y}|\mathbf{x}) \,. \tag{232}$$

Substituting this expression into the detailed balance condition we have

$$\frac{H(\mathbf{y}|\mathbf{x})}{H(\mathbf{x}|\mathbf{y})} = \frac{\pi(\mathbf{y})q(\mathbf{x}|\mathbf{y})}{\pi(\mathbf{x})q(\mathbf{y}|\mathbf{x})} \,. \tag{233}$$

Metropolis and Hastings suggested the choice

$$H(\mathbf{y}|\mathbf{x}) = \min\left(1, \frac{\pi(\mathbf{y})q(\mathbf{x}|\mathbf{y})}{\pi(\mathbf{x})q(\mathbf{y}|\mathbf{x})}\right) , \tag{234}$$

which automatically satisfies the detailed balance condition (233) since either $H(\mathbf{y}|\mathbf{x}) = 1$ and $H(\mathbf{x}|\mathbf{y}) = \pi(\mathbf{x})q(\mathbf{y}|\mathbf{x})/(\pi(\mathbf{y})q(\mathbf{x}|\mathbf{y}))$ or $H(\mathbf{x}|\mathbf{y}) = 1$ and $H(\mathbf{y}|\mathbf{x}) = \pi(\mathbf{y})q(\mathbf{x}|\mathbf{y})/(\pi(\mathbf{x})q(\mathbf{y}|\mathbf{x}))$. In our application we want to use the Metropolis–Hastings algorithm to generate the posterior distribution so we set $\pi(\mathbf{x}) = p(\mathbf{x}|\mathbf{d}, M)$ resulting in the acceptance probability for the state transition $\mathbf{x} \to \mathbf{y}$:

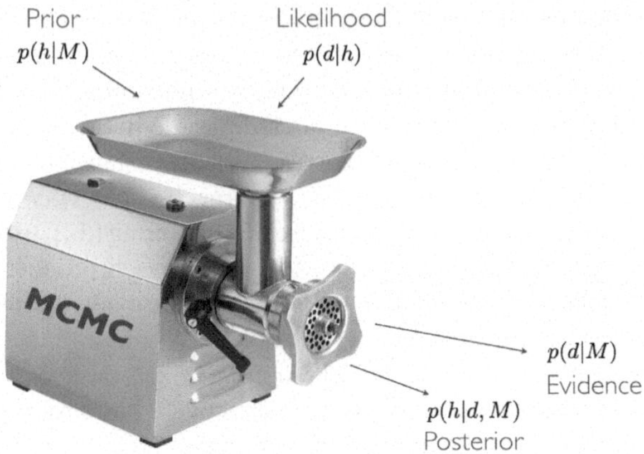

Fig. 42 How the sausage is made: the Metropolis–Hastings MCMC algorithm is a flexible approach for carrying out Bayesian inference

$$H(\mathbf{y}|\mathbf{x}) = \min\left(1, \frac{p(\mathbf{d}|, \mathbf{y}, M)p(\mathbf{y}, M)q(\mathbf{x}|\mathbf{y})}{p(\mathbf{d}|, \mathbf{x}, M)p(\mathbf{x}, M)q(\mathbf{y}|\mathbf{x})}\right). \tag{235}$$

Note that the evidence $p(\mathbf{d}, M)$ cancels in the Metropolis–Hastings (MH) ratio, so we only need the prior and likelihood, as illustrated in Fig. 42.

The MCMC algorithm proceeds by drawing some initial state $\mathbf{x}_{i=1} \sim p(\mathbf{x}, M)$ followed by the loop

- propose a new state $\mathbf{y} \sim q(\mathbf{y}|\mathbf{x}_i)$
- evaluate the MH ratio $H(\mathbf{y}|\mathbf{x}_i)$
- draw a random deviate $\alpha \sim U(0, 1)$
- if $H(\mathbf{y}|\mathbf{x}_i) > \alpha$ accept the new state, $\mathbf{x}_{i+1} = \mathbf{y}$, otherwise $\mathbf{x}_{i+1} = \mathbf{x}_i$
- increment $i \to i + 1$ and repeat

Assuming the process has converged, the collection of samples $\{\mathbf{x}_1, \mathbf{x}_2, \ldots\}$ generated by this algorithm represent fairs draws from the posterior distribution $p(\mathbf{x}|\mathbf{d}, M)$. The posterior samples can be used to estimate confidence intervals etc. It is often necessary to discard some number of samples from the beginning of the chain since it can take many iterations before the chain locks onto the region of high posterior density (known as the burn-in phase). With efficient proposal distributions the burn-in phase can be kept very short. The number of iterations needed depends on several factors. One factor is the degree of correlation between successive samples. The MH procedure generates correlated samples, and the degree of correlation can be measured by computing, for example, the auto-correlation length for each parameter. The number of independent samples can be estimated by dividing the total number of samples by the auto-correlation length of the most highly correlated parameter. But then there is the question of how many independent samples are needed, to which the answer depends on what you want to compute, and to what accuracy. For example, it takes many more samples to estimate a 95% credible region to 1% relative error than it does to estimate a 90% credible region to 10% relative error. The cost also increases with dimension, for example computing credible regions for the 2-d sky position of a source takes many more samples than computing the equivalent credible region for just the azimuthal angle.

The most important ingredient in a MCMC implementation is the proposal distribution. From (233) we see that the ideal proposal distribution would be the target distribution, $q(\mathbf{x}|\mathbf{y}) = \pi(\mathbf{x})$, since then $H(\mathbf{y}|\mathbf{x}) = 1$ and every proposed jump would be accepted, and each sample would be independent. But if we knew the target distribution (in our case the posterior distribution), and how to draw from it, there would be no need to perform the MCMC! In lieu of using the posterior distribution, we can instead compute approximations to the posterior distribution and use those as proposal distributions. For example, we can approximate the posterior distribution in the neighborhood of a local maximum using multivariate normal distributions with correlation matrices given by the inverse of the Fisher information matrix, as was done in Eqs. (208) and (209). To do this we need to locate maxima of the likelihood, which can be done using the algebraically maximized log likelihood employed in the searches (see Sect. 7.4), and either a grid search, or more efficient maximization schemes such as random re-start hill climbers, particle swarms, or genetic algorithms. Finding maxima of the likelihood surface can be computationally challenging, especially when the model dimension its high and/or the likelihood is expensive to compute, making it necessary to reduce the parameter dimension by ignoring less important parameters, and by using approximations to the likelihood function that are less expensive to compute. These approximate maps of the likelihood surface make for good global proposal distributions that can help the MCMC explore all the modes of a multi-modal posterior distribution.

The Fisher matrix approximation (208) also serves as a good local proposal distribution [13] as it takes into account correlations between parameters. To draw from the multi-variate normal distribution (208) we first find the eigenvalues e_i and associated eigenvectors $\mathbf{v}_i$ of the Fisher matrix, then propose jumps:

$$\mathbf{y} = \mathbf{x}_i + \frac{\beta}{\sqrt{e_j}} \mathbf{v}_j \,, \tag{236}$$

where $j \sim U[1, \dim(\mathbf{x})]$ and $\beta \sim \mathcal{N}(0, 1)$. The scaling by $1/\sqrt{e_j}$ yields a 68% acceptance rate if the Fisher matrix provides a faithful description of the posterior distribution. In practice the acceptance rate is lower due to the approximation being imperfect.

Another proposal distribution that is very effective at exploring parameter equations goes by the name *differential evolution* (DE) [6]. The procedure is very simple. First we collect some subset of N_h past samples from the Markov chain, called the history, $\{\mathbf{z}\}$, then propose a jump:

$$\mathbf{y} = \mathbf{x}_i + \gamma(\mathbf{z}_j - \mathbf{z}_k) \tag{237}$$

where $j, k \sim U[1, N_h]$ and $\gamma \sim \mathcal{N}(0, 1.68/\sqrt{\dim(\mathbf{x})})$. Here the scaling of the jumps, γ, is optimal for posteriors that follow a multi-variate normal distribution. The idea behind DE is that the vector $\mathbf{z}_j - \mathbf{z}_k$ connecting past samples provides a good guess

MCMC à la Montana

Ingredients:

Global likelihood maps

Fisher matrix proposals

Differential evolution proposals

Parallel tempering

Directions:

Mix all the proposals together. Check consistency by recovering the prior and producing diagonal PP plots. Results are ready when distributions are densely sampled and stationary.

Fig. 43 The standard MCMC recipe used by the Montana gravitational wave astronomy group

for the separation of future samples. In many applications DE has proven to be an incredibly effective proposal distribution, especially in situations where there are strong correlations between parameters. Some care has to be taken when employing DE as the use of past samples means that it is not strictly Markovian. The procedure can be shown to be asymptotically Markovian, meaning that if iterated long enough the samples will approach the stationary distribution. In practice this means having a sufficient number of independent posterior samples in the history. I typically keep 10^3 samples in a rolling history file by adding every 100th sample from the chain, and discarding the oldest sample from the history.

The final ingredient in my standard MCMC recipe, shown in Fig. 43, is to run multiple chains in parallel, and to allow exchanges between the chains. This procedure is variously called *Parallel Tempering* or *Replica Exchange*. The term *tempering* is taken from simulated tempering (also called simulated annealing), wherein the likelihood is flattened by raising it to a fractional power β, known as the inverse temperature. The terminology is borrowed from statistical mechanics and metallurgy, with the log likelihood playing the role of the energy. Each chain explores the annealed posterior distribution

$$p(\boldsymbol{\theta}|\mathbf{d}, \mathcal{H})_{\beta_i} = \frac{p(\mathbf{d}|\mathcal{H}, \boldsymbol{\theta})^{\beta_i}\, p(\boldsymbol{\theta}|\mathcal{H})}{p(\mathbf{d}|\mathcal{H})_{\beta_i}} . \tag{238}$$

Only samples from the $\beta_1 = 1$ "cold" chain can be used to define valid credible intervals etc, but the other chains serve an important purpose through parameter exchanges, and as a means to compute the model evidence. Periodically, swaps are proposed between the current state $\mathbf{x}_i$ of the β_i chain and the current state $\mathbf{x}_j$ of the β_j chain, and the swaps are accepted with probability

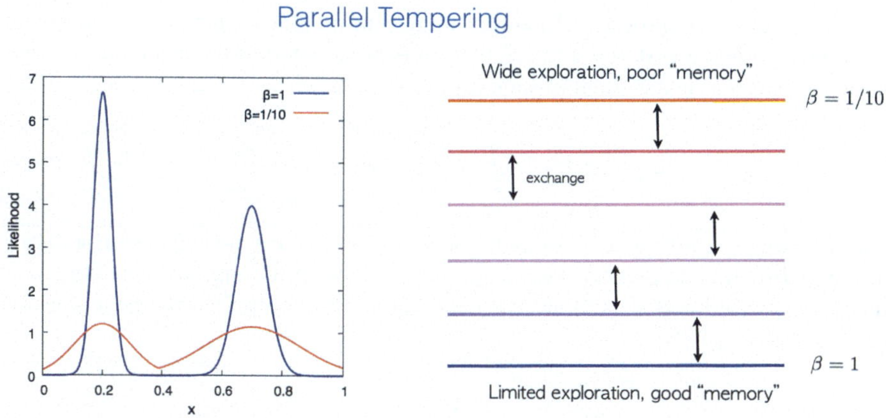

Fig. 44 Parallel tempering employs multiple chains on an inverse temperature ladder. Chains with higher temperatures (smaller β's) see a flatter likelihood surface (left panel) and are able to move more freely between local maxima. Exchanges between the chains (right panel) allow for the information to be shared and for good solutions to filter down to the cold chain, where the posterior samples are stored

$$H_{ij} = \min\left(1, \frac{p(\mathbf{d}|\mathcal{H}, \mathbf{x}_j)^{\beta_i}\, p(\mathbf{d}|\mathcal{H}, \mathbf{x}_i)^{\beta_j}}{p(\mathbf{d}|\mathcal{H}, \mathbf{x}_i)^{\beta_i}\, p(\mathbf{d}|\mathcal{H}, \mathbf{x}_j)^{\beta_j}}\right). \tag{239}$$

Note that only the likelihoods appear in the exchange probability. Parallel tempering is very effective at exploring multi-modal posteriors, as the hot chains explore a much flatter likelihood landscape that allows for free movement between local maxima, while the cold chains tend to lock onto high probability solutions and serve as "memory" for the ensemble (Fig. 44).

The spacing of the temperature ladder and the temperate range covered must be carefully chosen. A good rule of thumb is that the hottest chain should have $\beta \approx 1/\mathrm{SNR}^2$. The reasoning is that β factor re-scales the noise weighted inner product such that the effective signal-to-noise is $\mathrm{SNR}^2_\beta = \beta\, \mathrm{SNR}^2$, and we want $\mathrm{SNR}^2_\beta \approx 1$ for the hottest chain, rending the annealed likelihood sufficiently flat that the hot chain explores the full prior volume. The spacing of the chains has to be chosen such that chain exchanges are often accepted. If the chains are too widely spaced the inter-chain exchange probability gets very small, and the chains stop communicating. Conversely, if the chains are spaced too closely it takes a prohibitively large number of chains to cover the necessary temperature range. If the likelihood is well approximated by a multi-variate normal distribution it can be shown that the optimal spacing in geometric: $\beta_{i+1} = c\,\beta_i$ for some constant c. For the more complicated likelihood surfaces encounter in real-world analyses it is often necessary to use adaptive schemes to find the optimal placement of the temperature ladder. For simple examples a geometric spacing with $c = 0.8$ is usually a good choice, which for a $\mathrm{SNR} = 20$ signal requires $N_c = -2\log(\mathrm{SNR})/\log c = 27$ chains for full coverage.

A very useful by-product of parallel tempering is that it allows us to compute the model evidence $p(\mathbf{d}|\mathcal{H})$. The calculation is modeled after the calculation of the partition function in statistical mechanics:

$$\log p(\mathbf{d}|\mathcal{H}) = \int_0^1 \mathrm{E}[\log p(\mathbf{d}|\mathcal{H}, \mathbf{x})]_\beta \, d\beta \tag{240}$$

where the expectation value is computed with respect to the annealed posterior distribution $p(\boldsymbol{\theta}|\mathbf{d}, \mathcal{H})_\beta$. In practice we can approximate the above integral by the sum over the average likelihood at each inverse temperatures β_i, multiplied by the temperate spacing $\Delta\beta = \beta_{i+1} - \beta_i$.

7.6 *Worked Example—Sinusoidal Signal*

As a simple example, consider the signal from a slowly evolving binary system observed by a single LIGO-type detector. To keep the analysis tractable we will assume that the observation time $T_{\rm obs}$ is large compared to the orbital period $T = 2/f_0$, but small compared to the evolution chirp timescale $\tau = f/\dot{f}$. We will further assume that the observation time is short compared to timescale over which the detector moves so we can treat the detector as static. In other words, we are considering multiple cycles of a monochromatic signal with gravitational wave frequency f_0. Combining Eqs. (110) and (159) yields the detector response

$$\begin{aligned} h(t) &= \frac{2\mathcal{M}^{5/3}\omega^{2/3}}{R} \left(F_+(\boldsymbol{\Omega}, \psi)(1+\cos^2\iota)\cos(2\omega(t-t_0)+2\varphi_0)\right. \\ &\qquad \left. + F_\times(\boldsymbol{\Omega}, \psi)(2\cos\iota)\sin(2\omega(t-t_0)+2\varphi_0)\right) \\ &= A_0 \cos(2\pi f_0(t-t_0)+\phi_0) \end{aligned} \tag{241}$$

where $\omega = \pi f_0$,

$$A_0 = \frac{2\mathcal{M}^{5/3}\omega^{2/3}}{R} \sqrt{F_+^2(\boldsymbol{\Omega}, \psi)(1+\cos^2\iota)^2 + 4F_\times^2(\boldsymbol{\Omega}, \psi)\cos^2\iota}\,, \tag{242}$$

and

$$\phi_0 = \arctan\left(\frac{F_\times(\boldsymbol{\Omega}, \psi)(2\cos\iota)}{F_+(\boldsymbol{\Omega}, \psi)(1+\cos^2\iota)}\right) + 2\varphi_0\,. \tag{243}$$

With a single detector we are unable to separately measure the sky location $\boldsymbol{\Omega}$, distance R, inclination and polarization angles ι, ψ and initial orbital phase φ_0, but only combinations of these parameters that fix the overall amplitude A_0 and gravitational wave phase ϕ_0 seen in the detector. And absent a measurable chirp, τ, we are unable to measure the chirp mass $\mathcal{M}$.

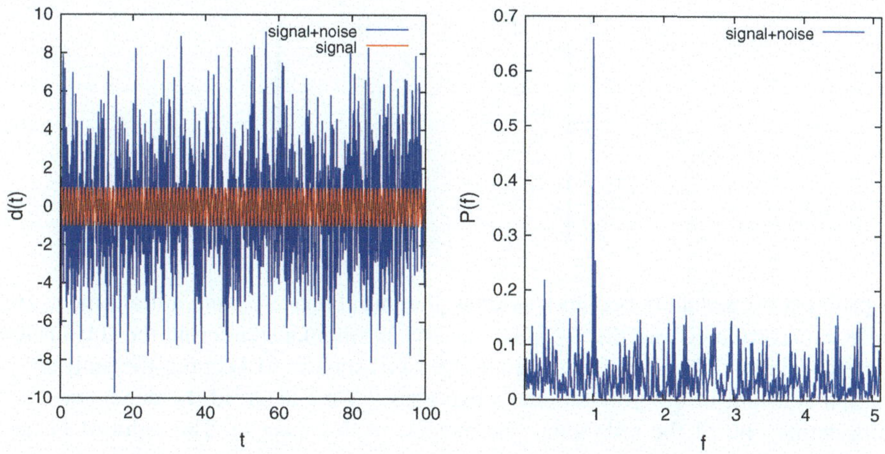

Fig. 45 Simulated data with a SNR = 7.07 sinusoidal signal embedded in white noise. On the left is the data in the time domain, and on the right is the power spectrum in the Fourier domain, where the signal is apparent as a spike at $f = 1$ Hz

Figure 45 shows simulated data generated with white Gaussian noise co-added to a signal with amplitude $A_0 = 1$, frequency $f_0 = 1$ and initial phase $\phi_0 = \pi$. The variance of the noise was adjusted to yield a match-filter signal-to-noise of SNR $= \sqrt{50} = 7.07$. While the signals looks to be buried in the noise when viewed in the time domain, it is readily apparent in the Fourier power spectrum.

Because the signal is monochromatic we can use Parseval's theorem to compute the SNR directly in the time domain:

$$\mathrm{SNR}^2 = (\mathbf{h}|\mathbf{h}) = 4\int \frac{\tilde{h}(f)\tilde{h}^*(f)}{S_n(f)}\,df = \frac{2}{S_n(f_0)}\int_0^{T_{\rm obs}} h^2(t)dt = \frac{A_0^2}{S_n(f_0)}\,T_{\rm obs}\,. \tag{244}$$

The elements of the Fisher information matrix can be evaluated in the same way as the SNR. Using the parameter set (A_0, f_0, ϕ_0, t_0) we find

$$\boldsymbol{\Gamma} = \mathrm{SNR}^2 \begin{pmatrix} \frac{1}{A_0^2} & \frac{1}{2f_0A_0} & 0 & 0 \\ \frac{1}{2f_0A_0} & \frac{4\pi^2T_{\rm obs}^2}{3} & \pi T_{\rm obs} & 2\pi^2 f T_{\rm obs} \\ 0 & \pi T_{\rm obs} & 1 & -2\pi f \\ 0 & 2\pi^2 f T_{\rm obs} & -2\pi f & 4\pi^2 f^2 \end{pmatrix} \tag{245}$$

Before attempting to estimate the parameter correlation matrix by inverting the Fisher matrix, it is instructive to look at the correlation matrix $\gamma_{ij} = \Gamma_{ij}/\sqrt{\Gamma_{ii}\Gamma_{jj}}$:

$$\gamma = \begin{pmatrix} 1 & \frac{\sqrt{3}}{4\pi f_0 T_{\rm obs}} & 0 & 0 \\ \frac{\sqrt{3}}{4\pi f_0 T_{\rm obs}} & 1 & \frac{\sqrt{3}}{2} & \frac{\sqrt{3}}{2} \\ 0 & \frac{\sqrt{3}}{2} & 1 & -1 \\ 0 & \frac{\sqrt{3}}{2} & -1 & 1 \end{pmatrix} \tag{246}$$

A problem immediately becomes evident: the initial phase ϕ_0 and initial time t_0 are fully anti-correlated and the ϕ_0, t_0 sub-matrix is singular, rendering the full Fisher matrix singular. Physically the degeneracy corresponds to keeping the sum $\phi_0 - 2\pi f_0 t_0$ constant in the gravitational wave phase. We can avoid the degeneracy by eliminating one of the redundant parameters, in this case t_0. The reduced Fisher matrix is then

$$\boldsymbol{\Gamma} = {\rm SNR}^2 \begin{pmatrix} \frac{1}{A_0^2} & -\frac{1}{2 f_0 A_0} & 0 \\ -\frac{1}{2 f_0 A_0} & \frac{4\pi^2 T_{\rm obs}^2}{3} & \pi T_{\rm obs} \\ 0 & \pi T_{\rm obs} & 1 \end{pmatrix} \tag{247}$$

with inverse

$$\boldsymbol{\Gamma}^{-1} \approx \frac{1}{{\rm SNR}^2} \begin{pmatrix} A_0^2 & 0 & 0 \\ 0 & \frac{3}{\pi^2 T_{\rm obs}^2} & \frac{3}{\pi T_{\rm obs}} \\ 0 & \frac{3}{\pi T_{\rm obs}} & 4 \end{pmatrix} \tag{248}$$

where we have used the fact that $f_0 T_{\rm obs} \gg 1$ to simplify the final expression. The diagonal elements of $\boldsymbol{\Gamma}^{-1}$ yield estimates for the 1-sigma parameter uncertainties:

$$\begin{aligned} \sigma_{A_0} &= \frac{A_0}{\rm SNR} \\ \sigma_{f_0} &= \frac{\sqrt{3}}{{\rm SNR}\, \pi T_{\rm obs}} \\ \sigma_{\phi_0} &= \frac{2}{\rm SNR}\,. \end{aligned} \tag{249}$$

Note that the parameter uncertainties all scale inversely with the SNR, which grows as the square root of the observation time. The error in the frequency decreases even more quickly with time. In general, quantities that impact the evolution of the phase are better constrained than those that impact the amplitude. The off-diagonal elements in $\boldsymbol{\Gamma}^{-1}$ tell us about the correlations in the parameter uncertainties.

Signal Search

The template metric (227) defines the line element

$$ds^2 = g_{\mu\nu} d\lambda^\mu d\lambda^\nu = \frac{4\pi^2 T_{\rm obs}^2}{3} df_0^2 + 2\pi T_{\rm obs} df_0 d\phi_0 + d\phi_0^2 \tag{250}$$

Maximizing with respect to the overall phase ϕ_0 using the procedure described in Sect. 7.3 reduces the search to just one parameter—the gravitational wave frequency f_0. The template spacing is then given by the line element

$$ds^2 = g'_{\mu\nu} d\lambda^\mu d\lambda^\nu = \frac{\pi^2 T_{\rm obs}^2}{3} df_0^2 \,. \tag{251}$$

The grid spacing to achieve a fitting factor FF is then

$$\Delta f_0 = \frac{2\sqrt{3}}{\pi T_{\rm obs}} \sqrt{1 - {\rm FF}} \,. \tag{252}$$

For example, setting $FF = 0.97$ yields a spacing of $\Delta f_0 \approx 0.2/T_{\rm obs}$. If want the search to cover signals that complete between 50 and 250 oscillations during the observation time, then the parameter volume is

$$V_{f_0} = \int_{50/T_{\rm obs}}^{250/T_{\rm obs}} \sqrt{g'}\, df_0 = \frac{200\,\pi}{\sqrt{3}} \,. \tag{253}$$

With $d = 1$ and $FF = 0.97$ the cell size is $\Delta V_{f_0} = 0.35$, and the number of templates in the bank is $N = V_{f_0}/\Delta V_{f_0} \approx 1000$. In this simple case with a one dimensional grid we can also compute the number of templates as the ratio of the search range $200/T_{\rm obs}$ divided by the template spacing Δf_0.

The probability distribution for the search statistic, $\Lambda_{\rm max}$, maximized over A_0, ϕ_0 and f_0, was determined empirically by repeating the search using 10^4 simulated noise realizations. The distributions are displayed in Fig. 46 for pure noise, and for noise co-added to a signal with SNR $= 7$. Setting a 1% false alarm rate yields a detection threshold of $\Lambda_{\rm max} = 22$, and a false dismissal probability of 0.5% for signals with SNR $= 7$.

The output of the grid search over f_0 for the simulated data set is shown in Fig. 47. The maximum likelihood template had $f_0 = 0.999$, $A_0 = 0.898$ and $\phi_0 = 2.75$. The match between this template and the injected signal was M $= 0.98$. The false alarm probability is too low to be reliably estimated from the empirically derived probability distribution for the noise hypothesis shown in Fig. 46, but the significance is greater than $3-\sigma$ (Gaussian equivalent, $p_{\rm FA} < 0.3\%$).

Parameter Estimation

The posterior distribution for the source parameters can be derived using the MCMC recipe described in Sect. 7.5 and illustrated in Fig. 43. Uniform priors were assumed for all parameters with ranges $A_0 \in [0, 10]$, $f_0 \in [0.5, 2.5]$ and $\phi_0 \in [0, 2\pi]$. A mix-

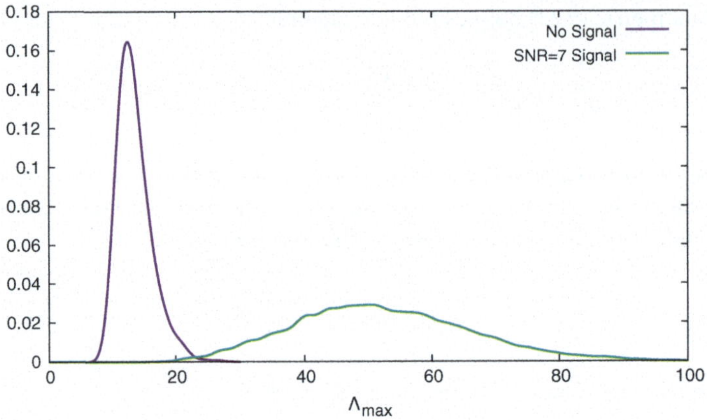

Fig. 46 Probability distributions for the $\Lambda_{\max}$-statistic in pure Gaussian noise, and for data containing a sinusoidal signal with SNR $= 7$

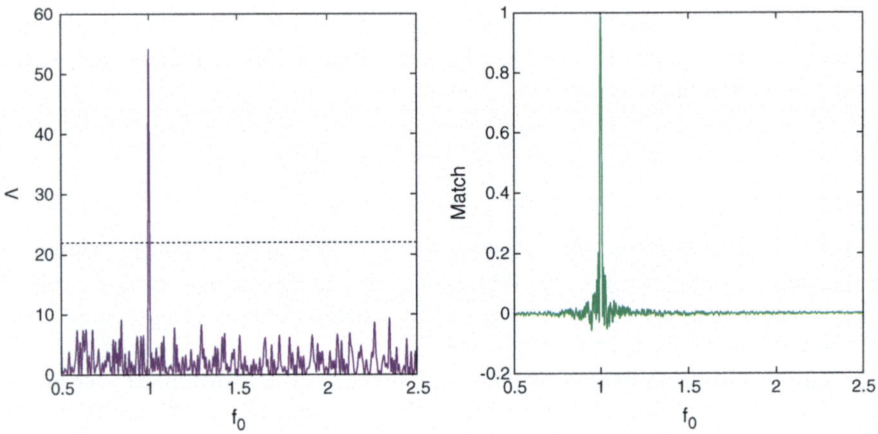

Fig. 47 The output of the grid search over f_0 for data containing a sinusoidal signal with SNR $= 7$. The panel on the left shows the matched filter statistic while the panel on the right shows the match between the best fit template at each frequency and the injected signal. The peak in the matched filter statistic exceeds the 1% false alarm probability detection threshold (shown as a dashed line)

ture of proposal distributions were used, made up of a global proposal, a multi-variate normal Fisher matrix using Eq. (247), and a differential evolution proposal. Parallel tempering was employed with 30 chains geometrically spaced by a factor of $c = 0.87$. The global proposal was constructed by normalizing the output of the matched filter search, $\Lambda_{\max}(f_0)$, shown in Fig. 47, and drawing f_0 from this distribution, while simultaneously drawing A_0 and ϕ_0 from their prior distributions. Note that we could have drawn A_0 and ϕ_0 from some distribution centered on their maximum likelihood values for each f_0, but uniform draws were sufficient for this simple example.

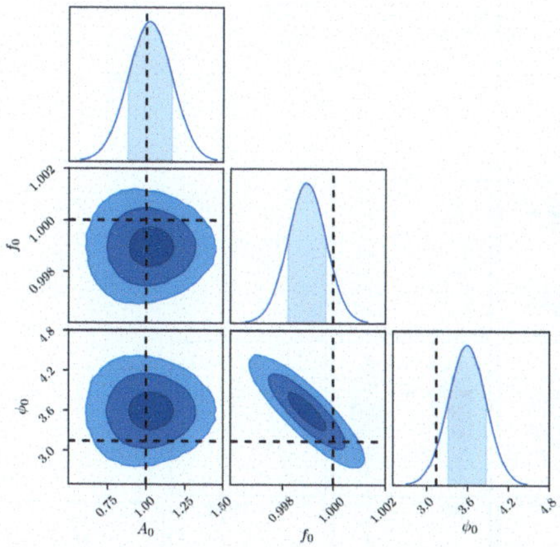

Fig. 48 The panel on the left displays a corner plot of the posterior distribution produced using the MCMC algorithm for the simulated data shown in Fig. 45 containing a SNR = 7.07 signal. The panel on the right shows the Fisher matrix approximation to the posterior distribution, centered on the maximum likelihood values for the parameters. Overall the agreement is good, but note that the 2-d probability contours for the MCMC derived posterior distributions are not perfectly elliptical

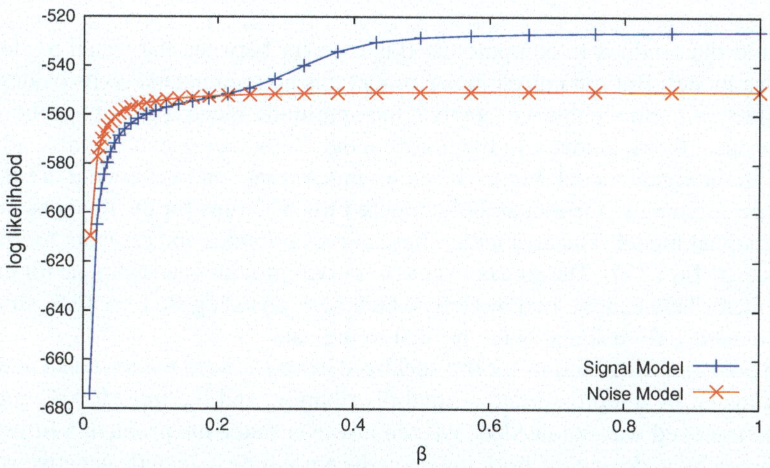

Fig. 49 The average log likelihood as a function of inverse temperature β for the signal model and the noise model. The area under these curves provides an estimate for the log of the model evidence. The area between the curves provides an estimate for the log Bayes factor

Figure 48 compares the posterior distribution derived by the MCMC algorithm to the predictions of the Fisher information matrix for the data shown in Fig. 45. The agreement is very good, but if you look closely you will see that the 2-d probability contours for the MCMC derived posterior distributions are not perfectly elliptical.

Since we are using parallel tempering it is also possible to compute the model evidence. On its own the evidence for the signal model is not very interesting. What

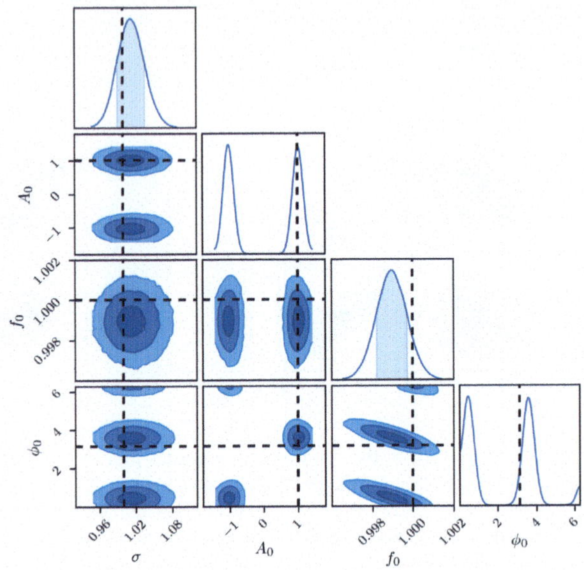

Fig. 50 The full multi-modal posterior distribution when the amplitude range is extended to negative values. The combination of the global proposal and parallel tempering allow the MCMC algorithm to fully explore both modes

we would like to do is to compute the Bayes factor between the signal model and the noise model. But out current noise model has no parameters, so its evidence is not defined. To remedy this we can treat the amplitude spectral density of the white noise, σ, as a free parameter in the noise model. To be consistent, we also allow σ to vary in the signal model. Figure 49 shows the average log likelihood as a function of inverse temperate β from parallel tempered MCMC runs for the the noise model and the signal model. The area under these curves provides and estimate for the log evidence via Eq. (240). The area between the curves provides and estimate for the log Bayes factor between the two models, which here gives $\log B_{S/N} = 14.1$, showing strong evidence for a signal being present in the data.

The posterior distributions for this idealized example were mono-modal and well approximated by a multi-variate normal distribution, and the full MCMC machinery we employed was not needed. We can however make the problem a little more challenging by widening the prior range on the amplitude to include negative values, $A_0 \in [-10, 10]$ which results in a multi-modal likelihood surface since solutions with parameters $(-A_0, \phi_0 + \pi)$ produce identical likelihoods to those with parameters (A_0, ϕ_0). Figure 50 shows the posterior distributions for the noise and signal parameters when the amplitude is allowed to take negative values. The combination of the global proposal and parallel tempering allow the MCMC algorithm to fully explore both modes. Without parallel tempering or the global proposal the chain remains stuck on a single mode of the posterior.

References

1. Abadie, J., et al.: A gravitational wave observatory operating beyond the quantum shot-noise limit: squeezed light in application. Nat. Phys. **7**, 962–965 (2011). arXiv:1109.2295 [quant-ph]
2. Abbott, B.P., et al.: Calibration of the advanced LIGO detectors for the discovery of the binary black-hole merger GW150914. Phys. Rev. D **95**, 062003 (2017). arXiv:1602.03845 [gr-qc]
3. Arzoumanian, Z., et al.: The NANOGrav 11-year data set: pulsar-timing constraints on the stochastic gravitational-wave background. Astrophys. J. **859**, 47 (2018). arXiv:1801.02617 [astro-ph.HE]
4. Audley, H., et al.: Laser interferometer space antenna (2017). arXiv:1702.00786 [astro-ph.IM]
5. Backer, D.C., Kulkarni, S.R., Heiles, C., Davis, M.M., Goss, W.M.: A millisecond pulsar. Nature **300**, 615–618 (1982)
6. Braak, C.J.F.T.: A Markov Chain Monte Carlo version of the genetic algorithm differential evolution: easy Bayesian computing for real parameter spaces. Stat. Comput. **16**, 239–249 (2006). https://doi.org/10.1007/s11222-006-8769-1. ISSN:1573-1375
7. Brooks, S., Gelman, A., Jones, G., Meng, X.: Handbook of Markov Chain Monte Carlo. CRC Press, Boca Raton (2011). https://books.google.com/books?id=qfRsAIKZ4rIC. ISBN:9781420079425
8. Brügmann, B.: Fundamentals of numerical relativity for gravitational wave sources. Science **361**, 366–371 (2018). https://science.sciencemag.org/content/361/6400/366.full.pdf. ISSN:0036-8075
9. Buonanno, A., Damour, T.: Effective one-body approach to general relativistic two-body dynamics. Phys. Rev. D **59**, 084006 (1999). arXiv:gr-qc/9811091 [gr-qc]
10. Buonanno, A., Sathyaprakash, B.S.: In: Ashtekar, A., Berger, B.K., Isenberg, J., MacCallum, M.E. (eds.) General Relativity and Gravitation: A Centennial Perspective, pp. 287–346. Cambridge University Press, Cambridge (2015)
11. Carroll, S., Carroll, S.: Spacetime and Geometry: An Introduction to General Relativity. Addison-Wesley, New York (2004). ISBN:9780805387322
12. Cornish, N.J.: Alternative derivation of the response of interferometric gravitational wave detectors. Phys. Rev. D **80**, 087101 (2009). arXiv:0910.4372 [gr-qc]
13. Cornish, N.J., Crowder, J.: LISA data analysis using MCMC methods. Phys. Rev. D **72**, 043005 (2005). arXiv:gr-qc/0506059 [gr-qc]
14. Cornish, N.J., Littenberg, T.B.: BayesWave: Bayesian inference for gravitational wave bursts and instrument glitches. Class. Quantum Gravity **32**, 135012 (2015). arXiv:1410.3835 [gr-qc]
15. Cornish, N.J., Romano, J.D.: Towards a unified treatment of gravitational-wave data analysis. Phys. Rev. D **87**, 122003 (2013). arXiv:1305.2934 [gr-qc]
16. Creighton, J., Anderson, W.: Gravitational-Wave Physics and Astronomy: An Introduction to Theory, Experiment and Data Analysis. Wiley, New York (2012). ISBN:9783527636044
17. Detweiler, S.: Pulsar timing measurements and the search for gravitational waves. Astrophys. J. **234**, 1100–1104 (1979)
18. Einstein, A.: Über das Relativitätsprinzip und die aus demselben gezogenen Folgerungen. (German) [On the relativity principle and the conclusions drawn from it]. Jahrbuch der Radioaktivität und Elektronik **4**, 411–462 (1908)
19. Einstein, A.: Die Feldgleichungen der Gravitation. Sitzungsberichte der Königlich Preußischen Akademie der Wissenschaften (Berlin), pp. 844–847 (1915)
20. Einstein, A., Infeld, L., Hoffmann, B.: The gravitational equations and the problem of motion. Ann. Math. (2) **39**, 65–100 (1938a)
21. Einstein, A., Infeld, L., Hoffmann, B.: The gravitational equations and the problems of motion. Ann. Math. (2) **41**, 455–564 (1938b)
22. Estabrook, F.B., Wahlquist, H.D.: Response of Doppler spacecraft tracking to gravitational radiation. Gen. Relativ. Gravit. **6**, 439–447 (1975)
23. Gelman, A., et al.: Bayesian Data Analysis. CRC Press, Boca Raton (2013). ISBN:9781439898208

24. Gilks, W., Richardson, S., Spiegelhalter, D.: Markov Chain Monte Carlo in Practice. Taylor & Francis, New York (1995). ISBN:9780412055515
25. Gralla, S.E., Wald, R.M.: A rigorous derivation of gravitational selfforce. Class. Quantum Gravity **25** [Erratum: Class. Quantum Gravity **28**, 159501 (2011)], 205009 (2008). arXiv:0806.3293 [gr-qc]
26. Hellings, R.W., Downs, G.S.: Upper limits on the isotropic gravitational radiation background from pulsar timing analysis. Astrophys. J. **265**, L39–L42 (1983)
27. Hewitson, M.: LISA science study team, LISA science requirements document, Issue 1.0 (2018). https://atrium.in2p3.fr/
28. Hobbs, G., Edwards, R., Manchester, R.: Tempo2, a new pulsar timing package. 1. Overview. Mon. Not. R. Astron. Soc. **369**, 655–672 (2006). arXiv:astro-ph/0603381 [astro-ph]
29. Isaacson, R.A.: Gravitational radiation in the limit of high frequency. II. Nonlinear terms and the effective stress tensor. Phys. Rev. **166**, 1272–1279 (1968b)
30. Isaacson, R.A.: Gravitational radiation in the limit of high frequency. I. The linear approximation and geometrical optics. Phys. Rev. **166**, 1263–1271 (1968a)
31. Izumi, K., Sigg, D.: Advanced LIGO: length sensing and control in a dual recycled interferometric gravitational wave antenna. Class. Quantum Gravity **34**, 015001. https://doi.org/10.1088/0264-9381/34/1/015001
32. Jackson, J.: Classical Electrodynamics. Wiley, New York (1975)
33. Kawamura, S., et al.: The Japanese space gravitational wave antenna: DECIGO. Class. Quantum Gravity **28**, 094011 (2011)
34. Kelley, L.Z., Blecha, L., Hernquist, L., Sesana, A., Taylor, S.R.: The gravitational wave background from massive black hole binaries in Illustris: spectral features and time to detection with pulsar timing arrays. Mon. Not. R. Astron. Soc. **471**, 4508–4526 (2017). arXiv:1702.02180 [astro-ph.HE]
35. Lehner, L., Pretorius, F.: Numerical relativity and astrophysics. Annu. Rev. Astron. Astrophys. **52**, 661–694 (2014). https://doi.org/10.1146/annurev-astro-081913-040031
36. Levin, J., Perez-Giz, G.: A periodic table for black hole orbits. Phys. Rev. D **77**, 103005 (2008). arXiv:0802.0459 [gr-qc]
37. Maggiore, M.: Gravitational Waves. Volume 1, Theory and Experiments. Oxford University Press, Oxford (2007). ISBN:9780191717666
38. Mathur, S.D.: What are fuzzballs, and do they have to behave as firewalls? In: Proceedings, 14th Marcel Grossmann Meeting on Recent Developments in Theoretical and Experimental General Relativity, Astrophysics, and Relativistic Field Theories (MG14) (In 4 Volumes): Rome, Italy, 12–18 July 2015, vol. 1, pp. 64–81 (2017)
39. Mino, Y., Sasaki, M., Tanaka, T.: Gravitational radiation reaction to a particle motion. Phys. Rev. D **55**, 3457–3476 (1997). https://link.aps.org/doi/10.1103/PhysRevD.55.3457
40. Misner, C., Thorne, K., Wheeler, J.: General Relativity. W. H. Freeman and Company, San Francisco (1968)
41. Owen, B.J.: Search templates for gravitational waves from inspiraling binaries: choice of template spacing. Phys. Rev. D **53**, 6749–6761 (1996). arXiv:gr-qc/9511032 [gr-qc]
42. Owen, B.J., Sathyaprakash, B.S.: Matched filtering of gravitational waves from inspiraling compact binaries: computational cost and template placement. Phys. Rev. D **60**, 022002 (1999). arXiv:gr-qc/9808076 [gr-qc]
43. Pais, A.: Subtle is the Lord: The Science and the Life of Albert Einstein. Oxford University Press, Oxford (2005). https://books.google.fr/books?id=0QYTDAAAQBAJ. ISBN:9780192806727
44. Poisson, E.: The motion of point particles in curved spacetime. Living Rev. Relativ. **7**, 6 (2004). https://doi.org/10.12942/lrr-2004-6. ISSN:1433-8351
45. Poisson, E., Will, C.: Gravity: Newtonian, Post-Newtonian, Relativistic. Cambridge University Press, Cambridge (2014). https://books.google.fr/books?id=PZ5cAwAAQBAJ. ISBN:9781107032866
46. Quinn, T.C., Wald, R.M.: Axiomatic approach to electromagnetic and gravitational radiation reaction of particles in curved spacetime. Phys. Rev. D **56**, 3381–3394 (1997). https://link.aps.org/doi/10.1103/PhysRevD.56.3381

47. Rakhmanov, M.: Fermi-normal, optical, and wave-synchronous coordinates for spacetime with a plane gravitational wave. Class. Quantum Gravity **31**, 085006 (2014). arXiv:1409.4648 [gr-qc]
48. Robson, T., Cornish, N., Liu, C.: The construction and use of LISA sensitivity curves. Class. Quantum Gravity **36**, 105011 (2019). arXiv:1803.01944 [astro-ph.HE]
49. Romano, J.D., Cornish, N.J.: Detection methods for stochastic gravitational-wave backgrounds: a unified treatment. Living Rev. Relativ. **20**, 2 (2017). arXiv:1608.06889 [gr-qc]
50. Schutz, B., Schutz, D.: A First Course in General Relativity. Cambridge University Press, Cambridge (1985). ISBN:9780521277037
51. Sivia, D., Skilling, J.: Data Analysis: A Bayesian Tutorial. Oxford University Press, Oxford (2006). ISBN:9780198568315
52. Skilling, J.: Nested sampling for general Bayesian computation. Bayesian Anal. **1**, 833–859 (2006). https://doi.org/10.1214/06-BA127
53. Van de Meent, M.: Modelling EMRIs with gravitational self-force: a status report. J. Phys.: Conf. Ser. **840**, 012022. https://doi.org/10.1088/1742-6596/840/1/012022

Supermassive Black Hole Accretion and Feedback

Andrew King

Contents

A. King (✉)
Department of Physics & Astronomy, University of Leicester, Leicester LE1 7RH, UK
e-mail: ark@astro.le.ac.uk
URL: https://www2.le.ac.uk/departments/physics/people/andrewking

Astronomical Institute Anton Pannekoek, University of Amsterdam, Science Park 904,1098 XH Amsterdam, The Netherlands

Leiden Observatory, Leiden University, Niels Bohrweg 2, 2333 CA Leiden, The Netherlands

R. Walter et al. (eds.), *Black Hole Formation and Growth*, Saas-Fee Advanced Course 48,
https://doi.org/10.1007/978-3-662-59799-6_2

Abstract I review the physics of accretion on to supermassive black holes in galaxy centres, and how this results in feedback affecting the host galaxy.

1 Introduction: Supermassive Black Holes in the Universe

Black holes are objects so compact that light cannot escape from them. A full understanding of them needs general relativity (GR), but we can often get useful insights by using Newtonian physics, while remembering two things: in relativity, light speed is the limiting velocity for all matter, and mass and energy are equivalent. Then we get an idea of just how compact a black hole is by equating the Newtonian escape velocity $v = (2GM/R)^{1/2}$ from an object of mass M and radius R to the speed of light c. This gives

$$R = R_s = \frac{2GM}{c^2} = 3 \times 10^{13} M_8 \,\text{cm}. \tag{1}$$

Here I have written $M_8 = M/10^8 \mathrm{M}_\odot$, to give a typical value for the sizes of the supermassive black holes which (as we shall see) inhabit the centres of most galaxies. From (1) this size is similar to the Earth–Sun distance. The quantity R_s is called the Schwarzschild radius. If instead I had taken a mass of order the Sun's mass, we would have found a Schwarzschild radius of order only 3 km: black holes are very compact. This is sometimes expressed as saying that they are very dense, but although this is true for stellar-mass black holes, whose mass densities $\sim\rho \sim 3M/4\pi R_s^3$ are even higher than the matter in an atomic nucleus, it is definitely not true for supermassive holes, whose mean density is comparable with that of water or even air.

Black holes are central to modern astrophysics for a simple reason: they give the most efficient way of getting energy from ordinary matter (i.e. other than matter-antimatter annihilation). In a Newtonian picture, matter of mass m falling radially from infinity towards a black hole formally reaches speed c at R_s. Then if it collides or interacts strongly with other matter it can give up kinetic energy of order 10% of its rest-mass energy mc^2. This is far greater than the nuclear fusion processes which power stars, which release at most about 0.7% of the total rest-mass energy of the matter fused in their centres, which itself is only a few percent of the total mass of the star.

This must mean that matter falling (or *accreting*) on to black holes powers the most energetic phenomena in the Universe at every scale. The brightest sources of all are quasars and active galactic nuclei (often collectively called AGN), whose luminosities can reach $\sim 10^{46}$–$10^{48}\,\mathrm{erg\,s^{-1}}$ or even more ($\sim 10^{13}$–$10^{15}\mathrm{L_\odot}$). These sources are far smaller and less massive than the galaxies they live in, so cannot be powered by stars—black hole accretion is the only possibility [44, 63]. Importantly, we see the dual nature of accretion: this both produces luminous electromagnetic emission, and simultaneously grows the masses (and as we shall see, may change the spins) of the black holes.

1.1 The Eddington Limit

Although accreting black holes can be the most powerful energy sources in the Universe, there is a limit on this power. The electromagnetic radiation resulting from accretion exerts pressure because the individual photons have momentum, and this interacts with the infalling matter. For simplicity we assume that this matter is pure hydrogen (in practice this is usually quite close to reality), and we further assume that the infalling matter has no net electrical charge. Then the infalling protons, carrying most of the gravitating mass, must be accompanied by an equal number of electrons. These need not be bound to the individual protons (neutral hydrogen) but must, as an ensemble, move with the protons (ionized hydrogen). Again for simplicity we assume that the accretion geometry is spherically symmetric, and leave to later the complications caused by deviations from this. Then each electron feels an outward radiation force

$$F_{\mathrm{rad}} = \frac{L\sigma_T}{4\pi r^2 c} \tag{2}$$

where r is the radial coordinate, L is the luminosity resulting from accretion, and σ_T is the Thomson cross-section for the scattering of photons by electrons. This outward force is opposed by gravity, which exerts a force

$$F_{\mathrm{grav}} = \frac{GM(m_p + m_e)}{r^2} \simeq \frac{GMm_p}{r^2} \tag{3}$$

on the electron and on the proton that must accompany it, where M is the mass of the accretor, and m_p, m_e are the proton and electron masses, and we have used the fact that $m_p >> m_e$ at the last step. We see that both forces vary as r^{-2}, so if one of them is bigger at any radius, it is bigger at all radii. Equating them shows that the quantity

$$L_{\rm Edd} = \frac{4\pi GMm_pc}{\sigma_T} = \frac{4\pi GMc}{\kappa} \simeq 1.3 \times 10^{38} M_0 \,{\rm erg\,s^{-1}} = 1.3 \times 10^{46} M_8 \,{\rm erg\,s^{-1}} \tag{4}$$

is a limiting value for the luminosity L, usually called the Eddington limit, specified purely by the gravitating mass M. Clearly if the luminosity exceeds this value, the accretion that is producing it would stop. In (4) we have defined the electron scattering opacity $\kappa = \sigma_T/m_p \simeq 0.34\,{\rm cm^2\,g^{-1}}$, and written M_0 for $M/{\rm M}_\odot$.

Although we have derived it in the context of accreting supermassive black holes (SMBH), it is obvious that $L_{\rm Edd}$ is a limit for luminosities derived in other ways, such as nuclear burning in stars. Indeed Eddington's own derivation of the result was for stars, and at the time (the 1920s) the energy source of stars was unknown. In practice, the luminosities of the brightest stars are very close to the values of $L_{\rm Edd}$ given by their masses. We might wonder whether the limit applies to realistic accreting systems, where we will argue that accretion is unlikely to be spherical. In practice it also seems to hold for both X-ray binaries and AGN, in line with simple analytic arguments [66]. We should note that this luminosity limit also means a limit on the rate at which a black hole can grow, as it specifies a critical accretion rate

$$\dot{M}_{\rm Edd} = \frac{L_{\rm Edd}}{\eta c^2}, \tag{5}$$

which would produce the Eddington luminosity (where η is the accretion efficiency, typically of order 0.1 for a black hole) and so presumably limits the rate at which matter can accrete.

1.2 AGN Spectra

AGN have distinctive spectra, which characteristically peak in the soft X-rays and Extreme Ultra-Violet light (EUV), but usually have significant emission in medium-energy X-rays (cf Fig. 1).

So we can regard X-ray emission as a sign of the mass growth of their SMBH. Because there are relatively few sources of such X-rays, we can turn this reasoning around (a trick astrophysicists use often). Cosmic X-ray detectors find nonzero fluxes even when not making observations pointed at specific sources, and we can identify this flux (the cosmic X-ray background) as X-rays emitted by black hole accretion somewhere in the nearby (low redshift) Universe. Using the typical AGN spectrum we can turn this into a lower limit on the total mass of the black holes in the low-redshift Universe, and so into an average mass of black hole per galaxy.

This is the Soltan argument [68], and tells us that the black hole 'ration' is a black hole mass $\sim 10^8 {\rm M}_\odot$ per medium-sized galaxy. Some of this might be distributed through galaxies in the form of millions of stellar-mass black holes that are (or once

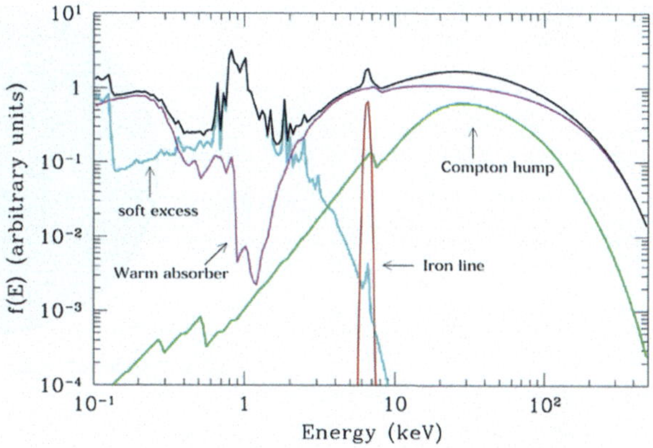

Fig. 1 A typical broadband spectrum of a quasar or active galactic nucleus

were) accreting gas from close companion stars in binary systems. But there is no observational support for this. Instead, the existence of AGN strongly suggests that much of the black hole ration is in the form of far larger supermassive holes. If these SMBH were only present in a minority of galaxies, their masses would have to be far higher—comparable with the galaxy masses themselves—for which there is no observational evidence. The important implication of this argument is that in the low-redshift Universe, essentially every galaxy must host a supermassive black hole. But since only a minority ($\lesssim 10^{-2}$) of low-redshift galaxies are active (i.e., have SMBH which accrete and so are detectable as AGN), accretion and black hole growth must occur only in shortlived phases, in every galaxy.

1.3 Where Are the Holes?

A large proportion of current work on SMBH aims to discover what determines when and for how long they accrete and grow. Although these questions remain unclear, the physical location of SMBH within their hosts is simple to understand. An SMBH of $\sim 10^8 M_\odot$ imprints a large perturbation on the motions of stars within the host galaxy. As the SMBH orbits within the host, its gravitational pull drags large numbers of stars behind it, raising a sort of wake like a boat on water. Just like a real wake, this exerts a drag force (see Fig. 2) on the moving SMBH.

This process is called dynamical friction [7], and is directly analogous to the effect of the Coulomb attraction of a charged particle moving through a plasma. Simple calculations show that the speed v of an SMBH of mass M moving through a galaxy with stellar mass density ρ obeys

$$\frac{\mathrm{d}v}{\mathrm{d}t} = -\frac{4\pi C G^2 M \rho}{v^2} \tag{6}$$

Fig. 2 Dynamical friction

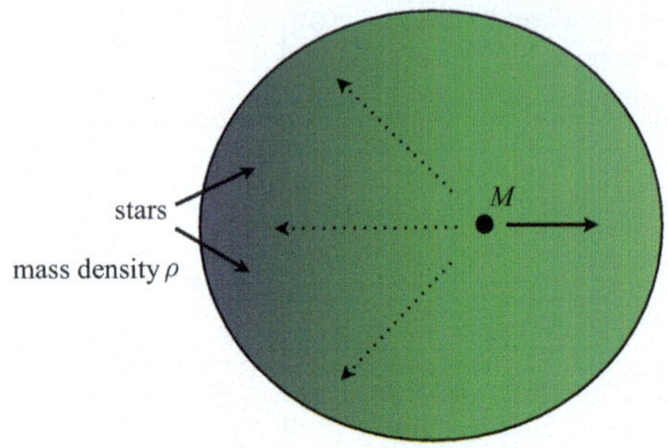

moving mass M slowed by raising gravitational 'wake' in star motions

where $C \simeq 10$ is a constant (the Coulomb logarithm, measuring the cumulative effect of the drag of many distant stars versus the bigger drag of the few nearby stars). Integrating, we have

$$v^3 = v_0^3 \left(1 - \frac{t}{t_{\rm fric}}\right) \tag{7}$$

where

$$t_{\rm fric} = \frac{v_0^3}{12\pi C G^2 M \rho} \tag{8}$$

where v_0 is the initial velocity. On the timescale $t_{\rm fric}$ the SMBH is reduced to rest, and must end up at the dynamical centre of the galaxy.

Now we can make a very crude estimate, modelling the central region of the galaxy as a uniform sphere of stars with total mass $M_g = 10^{11}{\rm M}_{\odot}$ and radius R_g = a few kpc, containing a $10^8{\rm M}_{\odot}$ SMBH moving with speed $v_0 \sim (2GM_g/R_g)^{1/2}$. This gives $t_{\rm fric} \lesssim 10^8$ yr, much smaller than the age $\sim 10^{10}$ yr of a low-redshift galaxy. So any SMBH is very likely to be in the centre of its host. If a given galaxy had more than one SMBH these are likely to have all rapidly collected in the centre and then merged under their mutual gravitational attraction. So we expect all but perhaps very small galaxies to have just one SMBH, living in its centre. By the same reasoning, a merger of two galaxies is likely to produce a single larger galaxy with a merged SMBH in the centre (Figs. 3 and 4).

There is now abundant observational evidence that this very simple one-to-one relation between galaxies and their central SMBH holds in the real Universe. We know a lot about the SMBH, known as Sgr A*, in the centre of our own galaxy. A long series of infrared observations by groups in Germany and the US show that the stars near Sgr A* move under its gravity precisely as expected from Newtonian physics and its weak-field general-relativistic extension, provided that its mass is about $4.5 \times 10^6{\rm M}_{\odot}$. See European Southern Observatory [14] for movies of the orbiting stars.

Fig. 3 A merging pair of galaxies

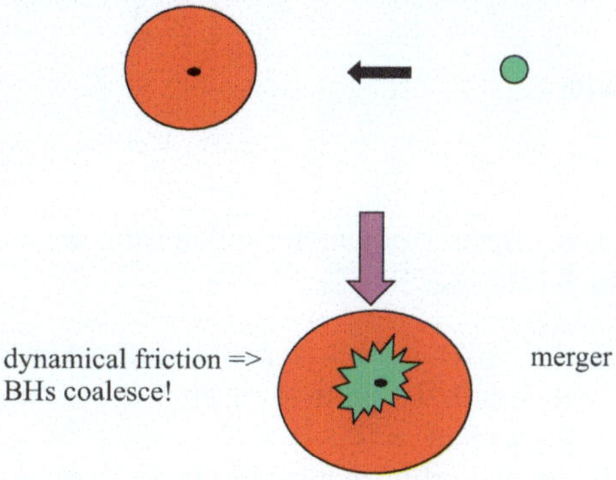

Fig. 4 The cosmological picture of galaxy growth: the large galaxy swallows a small one

2 Orbits Near Black Holes

Remarkably, all black holes in nature are described by a single family of solutions of the GR field equations (the Kerr solution), characterized uniquely by their masses M and angular momenta GM^2a/c, where a is a dimensionless constant with $0 \le a < 1$. Spinning black holes are rather more compact than nonrotating (Schwarzschild, $a = 0$) holes of the same gravitating mass M, because their rotational energy adds to their weight. It is usual to describe them using the gravitational radius

$$R_g = \frac{GM}{c^2} = \frac{1}{2}R_s, \tag{9}$$

rather than the Schwarzschild radius R_s.

In reality radial infall is not a very realistic possibility for an object as small as a black hole—matter almost always has significant angular momentum and so orbits the hole (put another way, matter is unlikely to be aimed precisely towards such a small object). So we should consider matter in orbit around a black hole. As pointed out above, we can use Newtonian mechanics to gain insight if we keep in mind the two relativistic features of limiting speed c and mass-energy equivalence. The energy equation for a particle in orbit around a Newtonian point mass is

$$\frac{v^2}{2} - \frac{GM}{r} = E \tag{10}$$

Now $v^2 = \dot{r}^2 + r^2\dot{\theta}^2$, where the angular velocity $\dot{\theta}$ is given by

$$r^2\dot{\theta} = J \tag{11}$$

where J is the specific angular momentum. This means we can write (10) as

$$\frac{1}{2}\dot{r}^2 + V(r) = E \tag{12}$$

with

$$V(r) = \frac{J^2}{2r^2} - \frac{GM}{r} \tag{13}$$

as an effective potential. In this form we can easily read off the familiar features of such orbits (cf Fig. 5):

1. if $J = 0$ the orbiting particle inevitably falls into the gravitating mass at $r = 0$
2. if $J \neq 0$, the particle escapes ($r \to \infty$) if $E > 0$, or remains bound (r remains finite) if $E < 0$
3. circular orbits require $\dot{r} = 0$, so $V(r) = E < 0 =$ constant, and $\mathrm{d}V/\mathrm{d}r = 0$, with $d^2V/\mathrm{d}r^2 > 0$ so that circular orbits are possible at minima of the effective potential $V(r)$.

We can see here the characteristic feature of all gravitationally bound systems, that particles orbiting with *lower* (i.e. more negative) total energy (here, at smaller r) have to move *faster*. This is the virial theorem: applied to stars made of hot gas it tells us that as they lose energy the gas gets hotter, which is the reason stars evolve rather than cooling down.

For motions near a black hole the important change is that we know speeds can become very large, but are not allowed to get bigger then c. Full relativity calculations of the geodesics in the Kerr metric show how the effective potential $V(r)$ has to change to accomodate this: it no longer has minima inside a certain radius. The radius of this innermost stable circular orbit (ISCO) depends on the Kerr angular momentum parameter a, and differs depending on whether the particle orbits in the

Fig. 5 The effective potential for circular orbits around a black hole

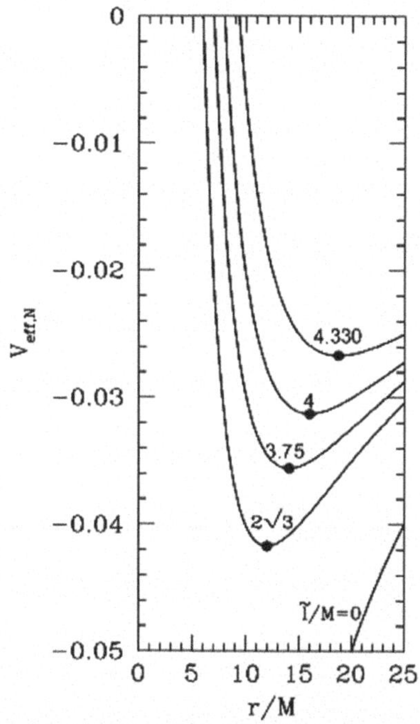

same sense as the black hole spin (prograde) or opposite to it (retrograde). Because it is deep in the gravitational potential of the black hole this orbit has low energy (and angular momentum) with respect to orbits further out—its total energy is more negative. So to have reached this tightly-bound orbit, the particle must have radiated this energy away, usually in electromagnetic form.

We see from Fig. 6 that the black hole spin a determines how much energy accreting matter releases when falling on a black hole. The figure shows that although accretion can have high energy efficiency $\eta \sim 0.4$, this requires high spin, with a approaching unity (and also that accretion is prograde). For rough estimates it is conventional to use $\eta \sim 0.1$, which formally holds for $a \sim 0.7$, and is within a factor 2 of the true value for all but extreme ($a \sim 1$) spin rates.

3 Supermassive Black Holes in Galaxies

Almost all supermassive black holes live in the central bulges of spiral galaxies, or in the centres of elliptical galaxies. In both cases these are near-spherical systems, which probably result from mergers with other galaxies. For a spiral, the mergers are minor, i.e. with smaller galaxies, while ellipticals are probably the outcome of a major merger with another spiral of a similar mass (Fig. 7).

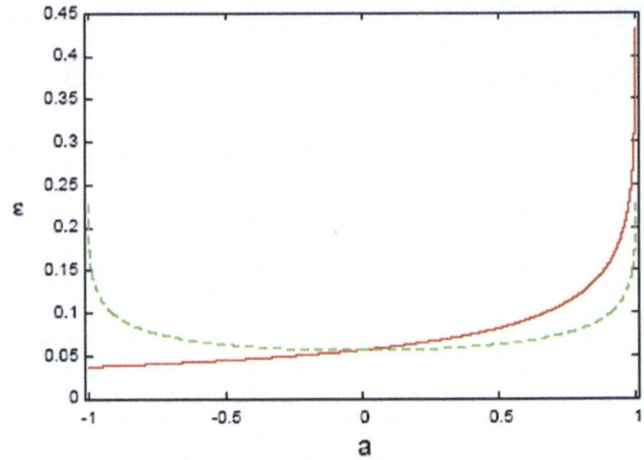

Fig. 6 The relation between black hole spin and accretion efficiency. The red curve gives the accretion efficiency as a function of spin parameter a. The green curve is a spin-symmetrized efficiency describing the (likely) case that accretion has a similar probability of being retrograde or prograde

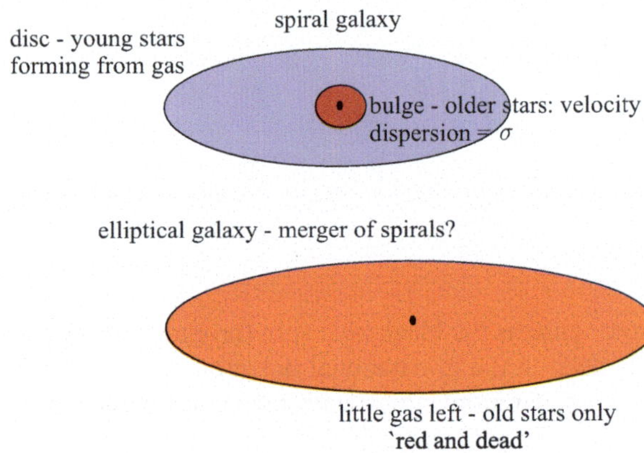

Fig. 7 Schematic picture of spiral and elliptical galaxies

In both cases, the stellar motions in the spheroidal component (bulge for a spiral, the whole galaxy for an elliptical) are generally characterised by velocity distributions which resemble the Maxwellian distribution of a hot gas, as expected theoretically [44]. In particular the velocity dispersions σ appear to be roughly spatially constant in any given system. In a hot gas this would imply a constant temperature, so these distributions are called *isothermal*. The corresponding mass distribution is

$$\rho = \frac{\sigma^2}{2\pi G r^2} \tag{14}$$

so that the total mass inside a radius R is

$$M_{\rm tot}(R) = 4\pi \int_0^R \rho(r) r^2 {\rm d}r = \frac{2\sigma^2 R}{G} \tag{15}$$

This means that the gravity of the SMBH alone controls motions only inside a region with $M(R) \lesssim M$, so within a radius

$$R_{\rm inf} = \frac{GM}{\sigma^2}. \tag{16}$$

called its sphere of influence. In a typical galaxy with $M \sim 3 \times 10^8 {\rm M}_\odot$, $\sigma = 200\,{\rm km\,s^{-1}}$ we have $R_{\rm inf} \simeq 24\,{\rm pc}$, so $R_{\rm inf}$ is far smaller than the typical scale $R_b \gtrsim 1$–$10\,{\rm kpc}$ of a spiral bulge, or the even larger scale of an elliptical galaxy.

Astronomers can now measure the masses M of a large sample of SMBH in low-redshift galaxies, generally by observing stars or gas within their spheres of influence (see Kormendy and Ho [41] for a review). When these are compared with the total mass M_b of the spheroid (bulge mass for a spiral, total galaxy mass for an elliptical) this gives the remarkable results shown in Fig. 8, which we can express as

$$M \sim 10^{-3} M_b, \tag{17}$$

and

$$M \simeq 3 \times 10^8 {\rm M}_\odot \sigma_{200}^{\alpha} \tag{18}$$

with $\sigma_{200} = \sigma/200\,{\rm km\,s^{-1}}$ and $\alpha \simeq 4.4 \pm 0.3$. The second relation can be compactly expressed as

$$M_8 \simeq 3\sigma_{200}^{\alpha}, \tag{19}$$

The discovery of these relations around the turn of the century transformed the study of supermassive black holes. These scalings show that the large-scale structure of galaxies 'knows' about the black hole mass, even though this has absolutely no direct gravitational influence on it, since $R_{\rm inf} << R_b$. Similarly it is not easy to see how the large-scale structure of the host spheroid could have a determining effect on the SMBH mass, which is controlled by conditions very close to it.

But there is one property of the SMBH which does potentially have a global significance for the host. We know from the Soltan relation (Sect. 1.2 above) that most of the SMBH mass was gained by luminous accretion. Then the $M-\sigma$ relation (18) shows that in reaching its current mass M, the hole has released an energy

$$E_{\rm BH} \sim \eta c^2 M \sim 2 \times 10^{61} M_8\,{\rm erg} \sim 6 \times 10^{61} \sigma_{200}^{\alpha}\,{\rm erg} \tag{20}$$

as radiation and possibly other (mechanical) forms. This is far larger than the binding energy

$$E_{\rm b} \sim M_b \sigma^2 \sim 8 \times 10^{58} \sigma_{200}^2\,{\rm erg} \tag{21}$$

which would be enough to disperse the entire spheroid. From this viewpoint, so far from being insignificant, the SMBH could have radically affected its host if only a small part of the released accretion energy was able to couple to it—a process called *feedback*. We will see later (Sect. 8.4) that it is likely that the growth of the SMBH

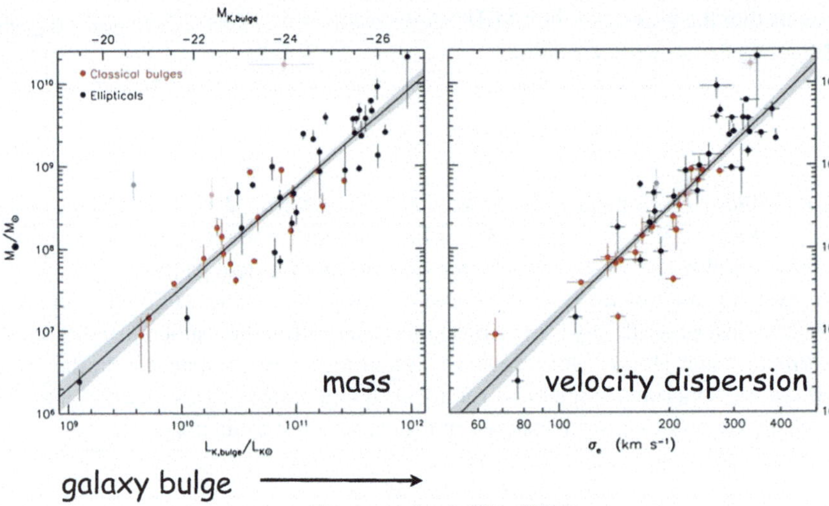

Fig. 8 The observed relation between supermassive black hole mass and the mass and velocity dispersion of the host galaxy spheroid. Galaxies know about central SMBH

mass is stopped once it reaches the mass specified by (18) because feedback from the SMBH disperses the gas that would otherwise drive further SMBH growth. But to understand this we need to study how black holes accrete and grow.

4 Accretion Discs

As we argued above, matter falling towards a supermassive black hole from the surrounding galaxy is extremely unlikely to be aimed directly at it. In other words it must have angular momentum about the hole. The closer to the SMBH the matter gets, the more important this angular momentum becomes in determining its motion. Ultimately it is in practice always enough to cause the gas to *orbit* the black hole, rather than fall radially into it. As the matter orbits, it is likely to lose energy in dissipative processes such as shocks. The gas always orbits at strongly supersonic speeds, so shocks occur wherever gas orbits intersect themselves or interact with neighboring matter. Shocks turn kinetic energy into heat, which is then radiated away, reducing the orbital energy. In contrast, there is little tendency for orbiting gas to change its angular momentum, as there are no strong torques acting on it (none at all if the only force is gravity). So the orbits of accreting gas always try to lose as much energy as is compatible with their angular momentum. It is a fairly obvious property of Newtonian (Kepler) orbits around a point mass that the orbit of lowest energy for a given angular momentum is a circle. So gas infall always results in a distribution of matter in circular orbits. If these result from infall of gas on similar initial orbits, the circles are likely to lie in a plane. We call this an *accretion disc*. Discs like this are a major theme of modern astrophysics, and for example are extremely common in stellar binaries, where matter transferred between two stars in orbit forms a disc in the binary orbital plane. I give an overview of the disc physics needed for studying SMBH accretion below, and refer for details to the book by Frank et al. [17], hereafter referred to as APIA, which gives an extensive discussion of standard disc theory.

We have already seen in Sect. 3 some of the properties of circular orbits, around both Newtonian point masses and black holes. We noted that for a black hole there is an innermost stable circular orbit (ISCO). Here direct infall inside the event horizon becomes inevitable, releasing the gravitational binding energy of the ISCO as electromagnetic radiation, and adding its angular momentum to the black hole. Clearly the first process is just what we need to power quasars and AGN, so the problem of understanding how these objects produce their observed luminosities reduces to ensuring that matter continues to reach the ISCO. This must involve some way of getting the gas at larger orbits in the disc gradually to give up some of its angular momentum and gently spiral inwards through a sequence of orbits which are almost circular. As the gas moves to smaller radii it releases some gravitational binding energy, and in many cases it is able to cool by radiating much of this away so efficiently that there is never a strong radial pressure gradient. Then the gas orbits are always closely similar to Keplerian circles, with circular velocity

$$v_K(R) = \left(\frac{GM}{R}\right)^{1/2} \tag{22}$$

at each radius R. The weakness of pressure forces compared to centrifugal ones is then equivalent to saying that the local sound speed c_s obeys $c_s << v_K$ at each radius. The residual pressure in the disc means that gas is supported slightly above and below the orbital plane, with a characteristic scaleheight H such that

$$\frac{H}{R} \simeq \frac{c_s}{v_K} << 1. \tag{23}$$

The three connected properties of efficient cooling, Keplerian rotation, and small scaleheight constitute the *thin disc* approximation [17], which frequently holds in practice (see Fig. 9).

But we still need a way of gradually removing angular momentum in order to drive gas inwards, that is, to make it accrete. In very general terms, the most likely way of achieving this exploits the fact that neighbouring orbits at slightly different radii *shear*: gas moves at slightly different rotational speeds, as in the thin-disc case where they have velocity $v_K(R)$. So any process giving a net transfer of rotational momentum across the interface gives a torque, usually described as *viscous* (cf Fig. 10).

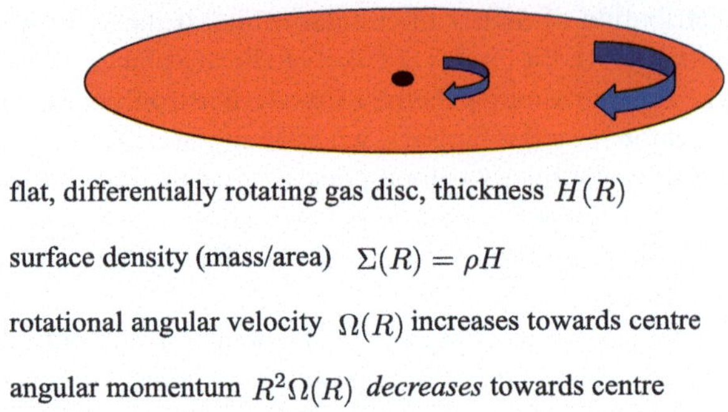

Fig. 9 The structure of a thin accretion disc

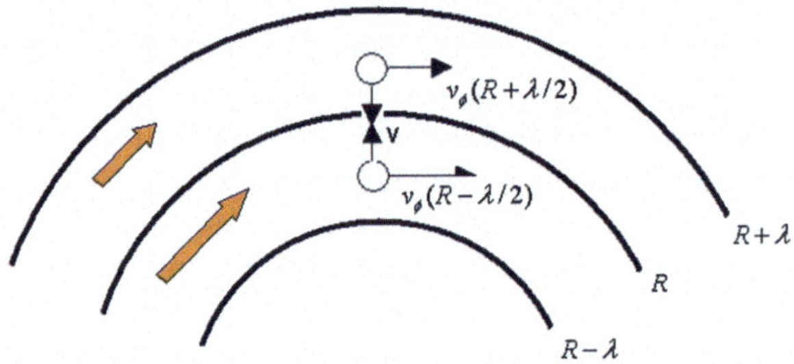

Fig. 10 Viscous torques in an accretion disc

This torque transports angular momentum between neighboring disc rings, so that the disc matter spreads both inwards and outwards in radius. If the viscous process carries the momentum at speeds v and deposits it at distances λ across the interface the resulting torque is proportional to λv. Standard 'molecular' viscosity does this using small-scale thermal motions (so λ is the mean free path, and v the sound speed) but something far stronger is needed to drive accretion through discs. We can parametrize the required value of the kinematic viscosity as

$$\nu = \alpha c_s H \tag{24}$$

Shakura and Sunyaev [66]. It is widely believed that the required mechanism giving ν must involve dynamo generation of magnetic fields within the disc plasma (specifically, the magnetorotational instability (MRI); Balbus and Hawley [3]), although this is still a matter of active research. In particular, current numerical simulations do not yet give values of α in agreement with observations of time-varying disc accretion [35].

The conservation equations for disc mass and angular momentum can be combined to give

$$\frac{\partial \Sigma}{\partial t} = \frac{3}{R}\frac{\partial}{\partial R}\left[R^{1/2}\frac{\partial}{\partial R}\left(R^{1/2}\nu\Sigma\right)\right] \tag{25}$$

where $\Sigma(R,t)$ is the disc surface density

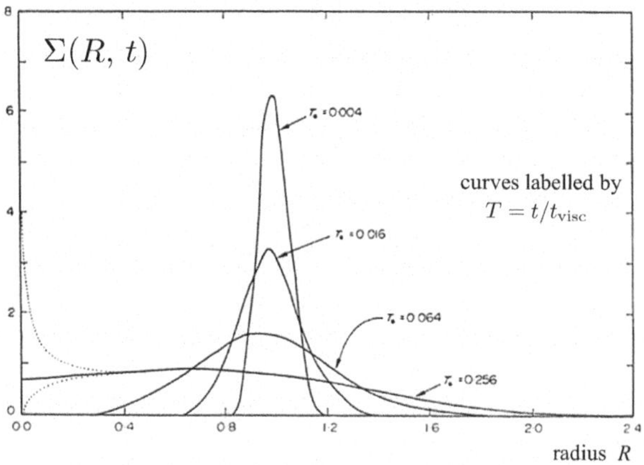

Fig. 11 The viscous spreading of a narrow ring. The initial ring spreads diffusely to make a disc

$$\Sigma = 2\pi \int_{-\infty}^{\infty} \rho(R, z, t) \mathrm{d}z \sim 2\bar{\rho}(R, t) H \tag{26}$$

where ρ is the mass density in the disc. Equation (25) shows that viscosity controls the time-dependence of the disc surface density through a *diffusion* equation, with the characteristic viscous timescale

$$t_{\text{visc}} = \frac{1}{\alpha} \left(\frac{R}{H}\right)^2 \left(\frac{R^3}{GM}\right)^{1/2} \tag{27}$$

directly related to local conditions at each disc radius. Figure 11 shows how an initial ring of matter spreads both to smaller and larger radii, to make a disc. Even the outward-moving gas eventually loses angular momentum to matter still further out and so ultimately spirals inwards—see Lynden-Bell and Pringle [45].

As well as controlling the mass flow, viscosity dissipates energy at the rate

$$D(R) = \frac{9}{8} \nu \Sigma \frac{GM}{R^3} \tag{28}$$

per unit area from each of the two disc surfaces (erg s^{-1} cm^{-2}). This equation relates the loss of gravitational binding energy ($\propto GM\Delta R/R^2$) as the disc matter spirals inwards a radial distance ΔR, to the rate the disc radiates energy away per unit surface area ($\propto 2\pi R \Delta R$) of the flat disc faces. This trade-off is not quite as simple as it may sound, because the viscous torques between the rings of gas also transport energy (see the discussion in Chap. 5 of APIA). In a steady state one can eliminate $\nu\Sigma$, and after some algebra we get

$$D(R) = \frac{3GM\dot{M}}{8\pi R^3} \left[1 - \left(\frac{R}{R_{\text{isco}}}\right)^{1/2}\right], \tag{29}$$

where $\dot{M}$ is the steady accretion rate, independently of ν. To complete the set of equations we need a relation specifying how the energy dissipation given by (28) is transported away from regions near the disc midplane to the flat radiating faces. For a disc which is optically thick over a local scaleheight H this is usually by radiative transport in the 'vertical' direction orthogonal to the disc plane, analogous to a one-dimensional star.

These simple equations have important implications for growing supermassive black holes. In particular it is obvious that there cannot be any mass growth involving accretion from parts of a disc whose viscous timescale is too long, certainly if it is longer than the age of the Universe at the epoch under consideration. At low redshift, this requirement is simply $t_{\rm visc} \lesssim t_{\rm H}$, where $t_{\rm H} \simeq 2 \times 10^{10}$ yr is the Hubble time. For accretion on to SMBH at significant rates we have $H/R \sim 10^{-3}$ [9]. With $\alpha \sim 0.1$, Eq. (27) gives the restriction

$$R_{\rm d} < 1 M_8^{1/3}\ {\rm pc}. \tag{30}$$

This tells us that the accretion which we know (from the Soltan argument) grows supermassive black holes must ultimately have so little angular momentum that it forms accretion discs smaller than the limit (30). In other words, the matter making up supermassive black holes *does* end up somehow almost 'aimed' at the hole with remarkable precision, although it cannot have had this property originally. It is still a matter of active research to discover why this is so. The lack of a clear candidate mechanism here means that we cannot yet expect to build a fully deterministic theory of SMBH growth from first principles. One suggestion [11] is that SMBH feeding is influenced by SMBH feed*back*, causing it to lose angular momentum, so that the SMBH 'forages' for its own food.

4.1 Self-gravity

I have so far tacitly assumed that the only gravity the disc feels is that of the accretor at its centre. But clearly it is possible for this assumption to fail, particularly in outer regions of the disc. Figure 12 shows that the gravity of local disc matter is stronger than that of the accretor if the vertical pull of a disc mass ρH^3 (where ρ is the local disc density, and H its scaleheight) is bigger than the vertical projection H/R of the accretor's gravity (its tidal field), i.e.

$$\frac{G\rho H^3}{H^2} > \frac{GMH}{R^3}, \tag{31}$$

or

$$\rho > \frac{M}{R^3}, \tag{32}$$

the usual relation one expects for tidal effects. We can also put this in the form of a limit on the total disc mass:

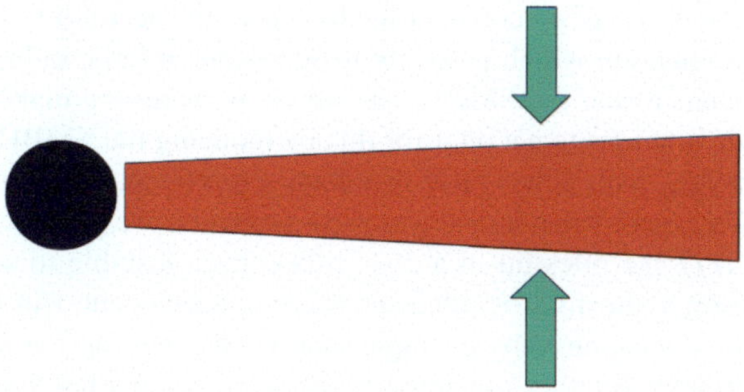

Fig. 12 Disc self-gravity

$$M_{\rm disc} \sim R^2 H \rho > \frac{H}{R} M \qquad (33)$$

These crude estimates agree with more sophisticated treatments (e.g. Toomre [72]). If these conditions hold, the usual conditions in accretion discs, particularly of rapid gas cooling, mean that the disc will locally try to fragment into self-gravitating clumps. If there is enough gas, these will be stars. The form of (33) shows that this is most likely in the outer part of any disc, and self-gravity is often the factor limiting the size of the disc. This is a plausible explanation for the ring of stars seen around the Milky Way's own SMBH, SgrA*. These may be the remains of the most recent major accretion event, which made a disc around the SMBH and whose outer part made stars. The disc has now been accreted, but the stars it formed inherited some of the disc angular momentum, and now orbit a little further out than the radius corresponding to the estimate (33) [33]. The self-gravity limit is important in several situations involving discs, as we shall see.

4.2 Spherical Accretion?

Given the difficulty of growing SMBH with matter which has angular momentum, it is tempting to avoid the problem altogether by assuming that the gas has somehow rid itself of this unwanted encumbrance. This kind of recipe is widely adopted in large-scale cosmological simulations of galaxy growth. These often assume that matter accretes steadily at a rate given by the Bondi [6] solution for accretion from a static, spherically symmetric medium with fixed mass density ρ_∞ and sound speed c_∞ at infinity:

$$\dot{M} = \dot{M}_{\rm B} \simeq 4\pi G^2 M^2 \frac{\rho_\infty}{c_\infty^3} \simeq \pi R_B^2 \rho_\infty c_\infty, \qquad (34)$$

where $R_B = 2GM/c_\infty^2$ is the Bondi radius. This has the virtue of producing a definite answer related simply to the surrounding conditions, but there are a number of reasons why this procedure is unsatisfactory. First, the assumption of *zero* angular momentum (wrt the accretor) is a singular limit: the importance of angular momentum grows like $\sim 1/r$ as we consider the flow near the accretor (cf Eq. (13)). So even an angular momentum which is totally insignificant at large radii comes to dominate the force balance near the SMBH—as we know, accreting matter tends to form a disc. There are attempts to account for this by reducing the SMBH accretion rate to some value $\beta^2 \dot{M}_{\rm B}$, with $\beta < 1$. But this cannot reproduce the qualitatively different behaviour we expect when a disc forms. In particular we will see that if enough mass $M_{\rm disc} \sim (H/R)M$ piles up in a disc, it becomes unstable to self-gravity effects: stars may form in the disc, or enhanced accretion can occur. But such strongly episodic effects are not captured by a simple steady-state assumption $\dot{M} \propto \dot{M}_{\rm B}$ for the SMBH mass growth. Another objection is the question of what was happening to the gas in the distant past, when the SMBH mass was very small. The assumed spherical symmetry means that the gas currently making up the SMBH was already exerting the same gravitational pull on distant matter at this time, so the assumption of a uniform density at infinity is questionable.

4.3 Steady Disc Accretion?

Another natural-looking assumption that leads to trouble is to suppose that the SMBH accretes from a steady accretion disc. The problem here is the disc's viscous timescale (27). As we saw, this is already comparable to the Hubble time for disc radii $R_{\rm d}$ only approaching $\sim$1 pc, far smaller than the scale of surrounding structures in the centre of a galaxy. But the steady-state assumption requires that a gas mass ΔM joining the disc at $R_{\rm d}$ leads *instantly* to the same gas mass ΔM accreting on to the central SMBH. The SMBH mass reacts to the newly-arriving gas without delay, even though in reality this may take a time comparable with the time on which the galaxy itself evolves.

4.4 Pitfalls

The last two subsections illustrate the maxim that when a calculation produces results that do not square with reality, it is rarely a result of a simple error in the algebra or the numerics, and much more often the consequence of an incorrect assumption made (often implicitly or unknowingly) before starting the calculation. A popular form of this is to assume an initial condition so unstable that nature would not have allowed the system to reach it. There is also a favoured remedy for this difficulty—cutting off the numerical calculation after one dynamical time, before the system has relaxed far enough from the initial condition to expose the problem.

4.5 Observations of Discs

There is by now a very large literature on accretion discs, and a body of observations that suggest that the simple picture sketched in this section provides a reasonable description of many known systems. Most of this evidence, and the relevant theory, is not from SMBH accretion at all, but from stellar-mass binary systems (see APIA for a review). Accretion is an almost unavoidable process for these objects, since stars can increase their radii markedly as they evolve, making it possible for the companion star to capture mass from it gravitationally. Only very wide systems with binary periods of tens of years are likely to finish their evolution without significant mass transfer between them. The mutual orbit of the two stars means that this transfer almost always involves an accretion disc. Further, if the star gaining matter from its companion is *compact* (a white dwarf, neutron star, or a stellar-mass black hole), accretion releases large luminosities which dominate over the intrinsic stellar luminosities deriving ultimately from nuclear burning. Studying systems like this (X-ray binaries) allowed astronomers to make the first well-grounded identification of black holes, partly because one can often estimate the velocity of one or both of the stars by using spectroscopy and the Doppler effect to give clean information about masses.

Because it is scale-free, we can directly apply the accretion disc theory we have discussed so far to the mass transfer process in these systems, and get insights that are useful for SMBH accretion. The most important results of this process concern disc timescales. There is very clear observational evidence (see Fig. 13) that the disc viscosity moves matter around on timescales given by $t_{\rm visc}$ (Eq. 27) with α not much less than unity—typically $\alpha \sim 0.3$–0.4. The light curves of outbursts from accretion discs which are thermally and viscously unstable show this very clearly: the characteristic long declines are close to exponential decays on the viscous timescale (Eq. 27), particularly in soft X-ray transients. Here the disc is self-irradiated by the central X-ray emission and so undergoes a pure viscous decay (see King and Pringle [34], cf Fig. 14).

As I mentioned earlier, this value of α deduced fairly directly from observation is currently something of a problem for current attempts to calculate disc viscosity from first principles using the magnetorotational instability (MRI), although there is still much work to do on the problem. Observations of SMBH discs have added relatively little to this comparison, mainly because AGN are more complex systems than close stellar-mass binaries, so interpreting observations is more difficult. But two things stand out: bright AGN were the first systems to show clear evidence for what is probably a universal feature, namely that systems close to their Eddington limits show observational evidence for strong outflows with velocities $v \sim 0.1c$. I discuss these in detail in Sect. 6.5. Second, AGN and especially radio galaxies very frequently show jets—matter in high-speed motion along apparently straight trajectories. These are often easily resolved, despite the large distances. We will see later that in general we expect the jet direction to be along the SMBH spin axis. It is now clear that many stellar-mass binaries also have jets, although these cannot be resolved spatially because of the very small angular scales.

some phenomena qualitatively independent of viscosity: only specifies *overall timescale* as

$$t_{\rm visc} \sim \frac{R^2}{\nu}$$

e.g. superhumps: requires orbital resonances within disc (Whitehurst & King, 1991; Lubow, 1991, 1992)

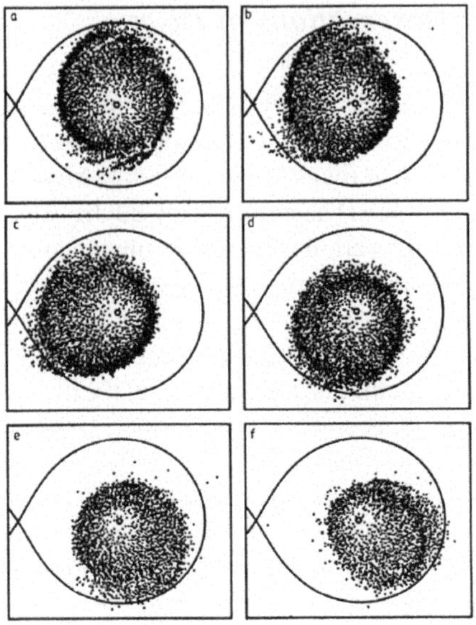

Fig. 13 Superhumps in an accretion disc

5 Misaligned Accretion

So far we have assumed that the discs we discuss are completely axisymmetric, so that in particular all the disc orbits lie in a single plane, and this plane itself is orthogonal to the black hole spin vector. This is a natural starting assumption to keep the theory simple. The first studies of discs were in the context of stellar-mass binary systems, where the binary orbital plane offers a preferred orientation for the disc. But even in this case there is no guarantee that the spin of an accreting stellar-mass black hole is orthogonal to this plane (cf King [24]): it appears that misaligned accretion is generic, and axisymmetry a singular limit. In an active galactic nucleus there is clearly no preferred plane for accretion—on the contrary, it appears very likely that flows reaching close enough to the SMBH for disc accretion to feed the hole may come from a variety of directions at various epochs. This type of picture—now often called chaotic accretion—was first considered by King and Pringle [33], and I will discuss it in more detail in the next section.

Misaligned accretion on to a black hole is subject to a general theorem by Hawking [19]: a stationary black hole is either static (has zero spin) or axisymmetric. The physical implication is that if a spinning black hole is in a non-axisymmetric environment, there must be torques acting to remove the non-axisymmetry. For a misaligned disc this means making the disc symmetrical about the black hole spin plane, so either aligned or counteraligned with the spin. The symmetrizing torques arise naturally: as disc material approaches the hole its orbits undergo Lense–Thirring precession (see Fig. 15). This precession is differential, i.e. its frequency changes

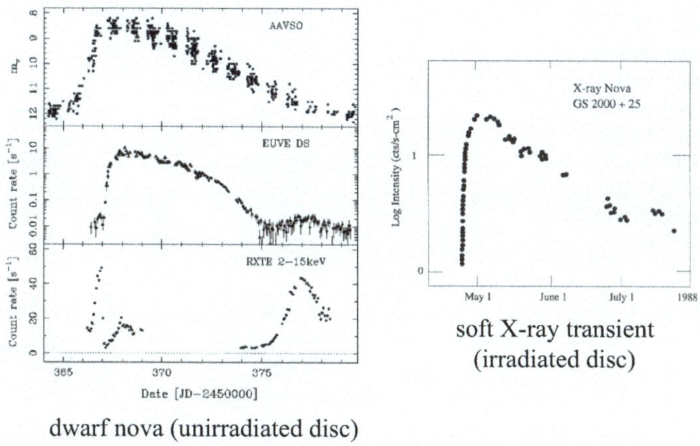

Fig. 14 Observed light curves from disc instabilities

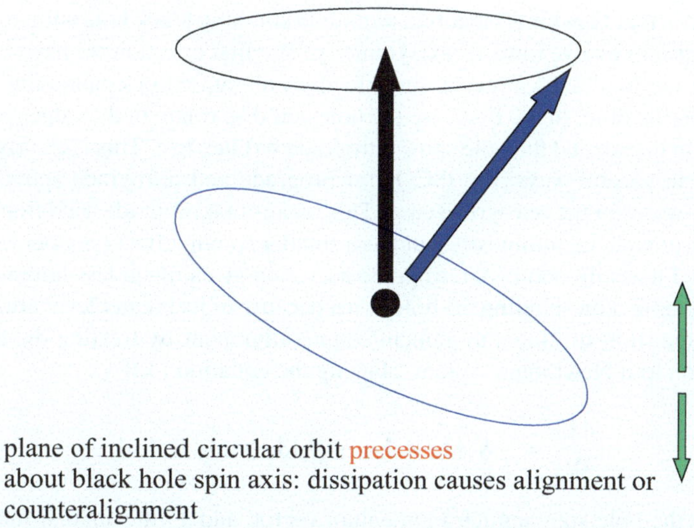

Fig. 15 The Lense–Thirring precession of inclined circular orbits around a black hole

with radius (as $\omega \sim R^{-3}$). So the disc is no longer plane, but warped, that is, its local plane changes with radius. The disc rings not only shear against neighbouring rings, but feel torques trying to dissipate the local up–down motions. Key to this is the way that viscosity adapts to a disc warp, where the forced Lense–Thirring precession produces a resonant radial pressure gradient (see Fig. 16).[1]

[1] Another example is if a spinning black hole is immersed in a misaligned uniform magnetic field, a torque acts to try to align the spin and the field, cf King and Lasota [31], King and Nixon [28].

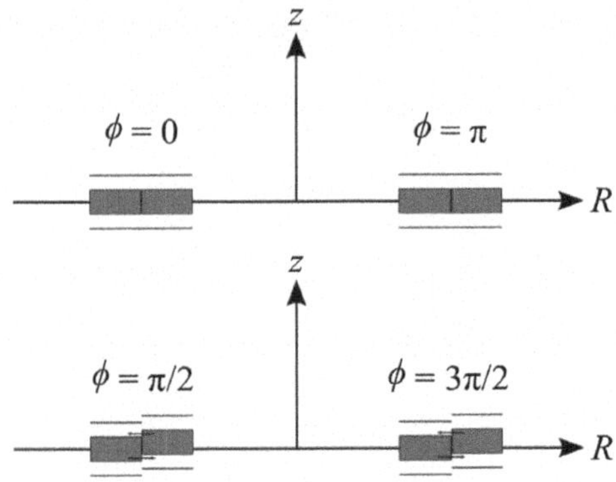

Fig. 16 A warped disc. The shaded areas have higher pressure. Arrows show pressure gradients induced by the warp. An orbiting fluid element feels a phase-dependent pressure gradient whose amplitude is a function of height

5.1 Alignment or Counteralignment?

We have seen that Hawking's theorem requires a spinning black hole with misaligned accretion disc to evolve towards axisymmetry. The first question we have to answer is whether the final axisymmetric state involves alignment or counteralignment of the disc and the hole. In the first case the hole and disc rotate in the same (prograde) sense, but in the second the hole spins retrograde wrt the disc. This is clearly a major difference in several ways: the ISCOs for prograde and retrograde spins are very different—wider in the retrograde case. This means that prograde accretion releases more electromagnetic luminosity, but has a smaller spinup effect (smaller lever arm) per accreted mass. In contrast, retrograde accretion gives much less luminosity, but has a bigger effect on spinning the hole *down* because of its greater lever arm. We can study the question of global alignment/counteralignment by treating the hole-disc combination as a Newtonian system, obeying the equation [32]

$$\frac{d\mathbf{J}_h}{dt} = -K_1[\mathbf{J}_h \times \mathbf{J}_d] - K_2[\mathbf{J}_h \times (\mathbf{J}_h \times \mathbf{J}_d)]. \quad (35)$$

Here $\mathbf{J}_h$ is the hole spin angular momentum vector, and $\mathbf{J}_d$ the angular momentum of the disc. Strictly speaking, it is the angular momentum of that part of the disc which feels a significant alignment torque, e.g. where the disc orbits have completed at least one Lense–Thirring precession. The other way of saying this is that $\mathbf{J_d}$ is the angular momentum of that part of the disc currently tending to warp under the alignment torque. Over time this radius increases to include more and more of the disc—the warp moves outwards—so in principle the magnitude J_d grows in time. But alignment/counteralignment is faster than this growth (by definition), so in practice we can regard J_d (but not $\mathbf{J_d}$!) as constant.

The degree of misalignment of disc and hole spin is measured by the vector quantity $\mathbf{J}_h \times \mathbf{J}_d$, so all the alignment torques must depend on it. Equation (35), where K_1, K_2 are functions of the disc structure, is then the most general form possible if the hole feels only precessions. The torques on the rhs are orthogonal to $\mathbf{J}_h$, where the first term on the rhs describes the torque inducing precession, and the second one describes alignment or counteralignment. I will show below that the quantity K_2 must be positive if there is dissipation, as we expect, since viscosity is involved in the alignment/counteralignment process.

Taking the scalar product of Eq. (35) with $\mathbf{J}_h$ shows that $\mathrm{d}J_h^2/\mathrm{dt} = 0$, that is, the magnitude of the spin remains constant during precession and alignment/ counteralignment. Thus the tip of the vector $\mathbf{J}_h$ moves on a sphere. The total angular momentum $\mathbf{J}_t = \mathbf{J}_h + \mathbf{J}_d$ is of course a constant vector, representing a fixed direction in space. Using this, and the fact that $\mathbf{J}_h.\mathrm{d}\mathbf{J}_h/\mathrm{d}t = 0$, we see that

$$\frac{\mathrm{d}}{\mathrm{d}t}(\mathbf{J}_h.\mathbf{J}_t) = \mathbf{J}_t.\frac{\mathrm{d}\mathbf{J_h}}{\mathrm{d}t} = \mathbf{J}_d.\frac{\mathrm{d}\mathbf{J}_h}{\mathrm{d}t}. \tag{36}$$

Now from Eq. (35) we get

$$\frac{\mathrm{d}}{\mathrm{d}t}(\mathbf{J}_h.\mathbf{J_t}) = \frac{\mathrm{d}}{\mathrm{d}t}(J_h J_t \cos\theta_h) = K_2[J_d^2 J_h^2 - (\mathbf{J}_d.\mathbf{J}_h)^2] = J_d^2 J_h^2 \sin^2\theta \equiv A > 0. \tag{37}$$

where θ is the angle between $\mathbf{J_h}$ and $\mathbf{J_d}$. Since both J_h and $\mathbf{J_t}$ are constant this means that

$$\frac{\mathrm{d}}{\mathrm{d}t}(\cos\theta_h) > 0 \tag{38}$$

where θ_h is the angle between the spin direction $\mathbf{J_h}$ and the fixed direction $\mathbf{J_t}$. So θ_h always decreases, meaning that the spin vector tends to align itself along the fixed total angular momentum (a.m.) vector $\mathbf{J_h}$. It is easy to show also [33] that

$$\frac{\mathrm{d}}{\mathrm{d}t}J_d^2 = -2A < 0 \tag{39}$$

We see that the magnitude of the disc angular momentum $J_d{}^2$ decreases as $\mathbf{J_h}$ aligns with $\mathbf{J_t}$ if and only if $K_2 > 0$, justifying the assumption made above. Although the total angular momentum of the system (hole plus disc) is conserved, alignment works because of dissipation. The magnitude of the spin of the hole remains unchanged, so dissipation must reduce the magnitude of the disc angular momentum.

We can now easily deduce the criterion for co- or counter-alignment. The vectors $\mathbf{J_h}$, $\mathbf{J_t}$ and $\mathbf{J_d}$ make a triangle. As co- or counteralignment proceeds the first two vectors have constant length, while the third shortens. Once $\mathbf{J_h}$ lines up with $\mathbf{J_t}$, it will be counteraligned wrt $\mathbf{J_d}$ if and only if

$$J_h^2 > J_t^2, \tag{40}$$

i.e. the hole spin vector is *longer* than the total a.m. vector, and so the disc a.m. is has to be counteraligned to the spin.

So given an initial angle θ between $\mathbf{J_h}$ and $\mathbf{J_d}$, the cosine theorem gives

$$J_t^2 = J_h^2 + J_d^2 + 2J_h J_d \cos\theta \tag{41}$$

Then counteralignment ($\theta \longrightarrow \pi$) occurs if and if the initial angle obeys

$$\cos\theta < -\frac{J_d}{2J_h}. \tag{42}$$

So counteralignment is a possible outcome, and requires both

$$\theta > \pi/2, \quad J_d < 2J_h. \tag{43}$$

—see Fig. 17.

We see that there is a cone of directions for the initial disc a.m. in which counter-alignment is the outcome. The process is subtle, in that it is not necessarily monotonic [32]—the spin and disc angular momenta can move initially towards one outcome, but in the end reach the opposite one.

The first treatment of this process Scheuer and Feiler [65] actually concluded that the only possible outcome was alignment of disc and spin, in flat contradiction of what we have just worked out. This erroneous result followed from an unsound assumption (cf the discussion in Sect. 4.3!), namely that for convenience, the authors decided to assume a Newtonian rest frame in which one of the vectors $\mathbf{J_d}, \mathbf{J_h}$ was itself at rest. Their choice was $\mathbf{J_d}$. But in reality only the total angular momentum $\mathbf{J_d} + \mathbf{J_h}$ is fixed in a Newtonian sense, so this assumption is only compatible with the conservation of angular momentum if J_d is *infinite*. Then from Eq. (42) we see that the conditions for counteralignment can never be met, as $\cos\theta$ has to be infinitely negative. This result—initially unchallenged—has had seriously misleading consequences for the subject, as we will see in the next section. But first we examine the effects of warping more closely.

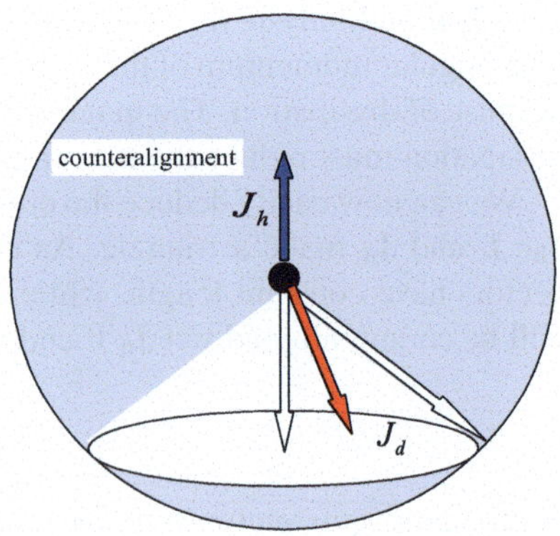

Fig. 17 Aligning accretion discs and spinning black holes: counter-alignment occurs for disc angular momenta within the white cone

5.2 Disc Warping, Breaking and Tearing

We have seen the inner part of the disc, where Lense–Thirring precession is rapid, aligns fastest with the spin plane, and so a disc *warp* travels outwards. In many cases, if the disc is steadily fed matter misaligned wrt the spin this results in a smooth steady warp, a result first found by Bardeen and Petterson [4] (Fig. 18).

For a time there was a general belief that this was the only possible outcome. But the behaviour of the viscosity in a warped disc is more complex than in Bardeen and Petterson [4], which simply assumes that there is a second independent viscosity coefficient α_2 relating to 'vertical' shearing of disc annuli. But if one instead makes the reasonable assumption that *dissipation* by viscosity is isotropic (cf Papaloizou and Pringle [54]) this enforces a relation of the form

$$\alpha_1 \alpha_2 = \frac{2}{\alpha^2} \tag{44}$$

where $\alpha_1 \simeq \alpha$ is the 'standard' Shakura–Sunyaev coefficient defining the isotropic dissipation. The key to understanding this is that in a warped disc, neighbouring annuli exert resonant torques on each other, so that the effects of dissipative damping work in a way that at first appears counter-intuitive. As we can see from (44), a

assumed warp (Lodato & Price 2010)

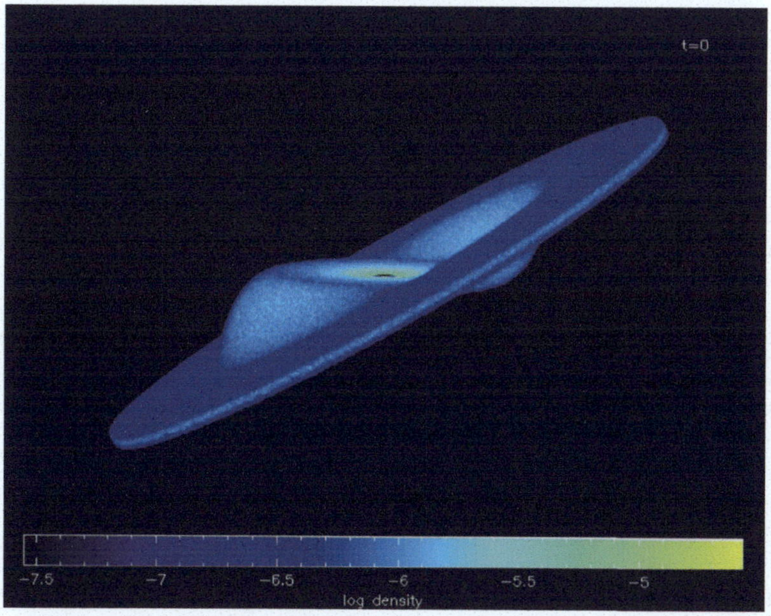

Fig. 18 A strong warp in an accretion disc with significant viscosity

smaller α makes α_2 increase strongly, so that annuli are pulled more strongly in the vertical direction than in the usual disc plane. In effect the forces trying to hold the disc together actually become weaker as the disc warps. For a sufficiently large warp amplitude it is possible for the disc to *break*—undergo an almost discontinuous change of plane. Matter accretes from the outer to the inner disc through a sequence of orbits whose inclination changes very sharply at a very narrow range of radii. This effect was found first by Lodato and Price [43], who assumed an initial disc which was already strongly inclined, but it also results self-consistently from misaligned time-dependent accretion [52].

By itself, disc breaking is not perhaps a big change from the original Bardeen–Petterson smooth warp. But the forced Lense–Thirring precession is still operating, and its rate is strongly dependent on disc radius (frequency $\omega \propto r^{-3}$). For sufficiently high inclination, rings of gas can break off from the misaligned outer disc, and through differential precession end up in partial *counterrotation*. As they spread viscously they cancel angular momentum with neighbouring disc annuli rotating closer to prograde, causing gas to fall to smaller radii, where the whole process can repeat itself as the infalling matter circularises. The net result is that disc matter has 'borrowed' angular momentum from the black hole via the Lense–Thirring effect in order to cancel its own. This process is called disc tearing [51]. It is clear that tearing can alter the instantaneous accretion rate on to the hole, but not of course its long-term average, which is set by the viscous timescale of the outer disc. It may also be possible to observe tearing directly, because of the large infall velocities (fractions of c) and corresponding large (but transient) redshifts it causes [58].

6 Chaotic Accretion and Supermassive Black Hole Growth

I introduced the idea of chaotic accretion above. It is inherently plausible simply because the physical scale of the SMBH is so much smaller than the central bulge of the host galaxy where it lives. But there is direct observational evidence favouring it also. As mentioned in Sect. 4.4, we expect that jets are directed along the SMBH spin axis. Observation (e.g. Kinney et al. [40]) is unambiguous that there is no correlation between observed jet directions and the large-scale structure of the host galaxy (see Figs. 19 and 20). This strongly suggests that the matter accreting on to the SMBH cannot have any memory of the large-scale structure of its host galaxy, and so is chaotic.[2]

6.1 Supermassive Black Hole Growth

Chaotic accretion is important in discussing one of the biggest constraints on models of black hole growth. Observation of quasars at high redshift give lower limits on

[2]This description—suggested by Piero Madau—is sometimes challenged on the grounds that mathematically the accretion process is stochastic, rather than chaotic. But the original name has stuck.

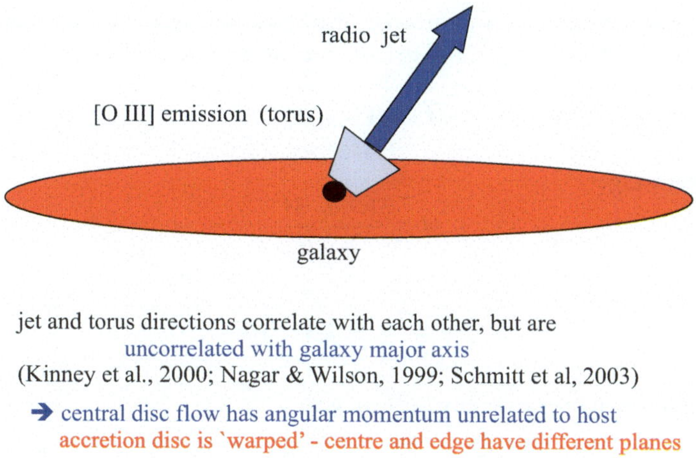

Fig. 19 The geometry of active galactic nuclei

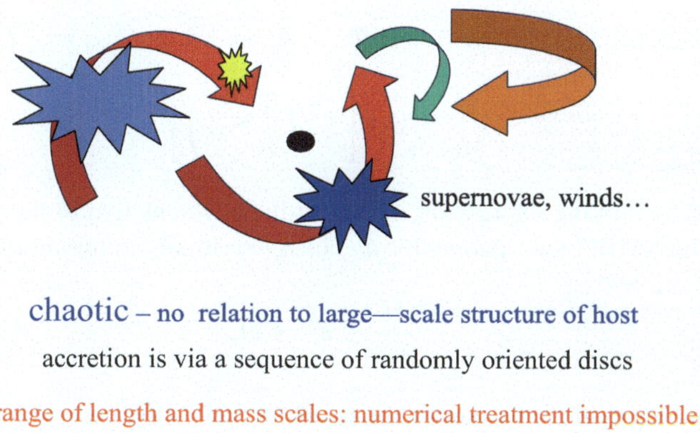

Fig. 20 Chaotic accretion: gas flows in the central regions of an active galaxy are probably not smooth and ordered

SMBH masses if we assume that their luminosities cannot significantly exceed the Eddington limit. The quasar cosmological redshifts give the look-back time since the Big Bang, and so an upper limit to the time that these SMBH can have been accreting. The results are challenging for accretion theory: there are SMBH with masses $M > 5 \times 10^9 \mathrm{M}_\odot$ [5, 74] at redshift $z = 6$, only 10^9 yr after the Big Bang.

The high masses observed at $z = 6$ make it natural to ask what initial masses these holes must have started with at high redshift. The Soltan relation tells us that SMBH gained most of their mass via luminous accretion. This probably means that they cannot have accreted mass any faster than at the rate giving the Eddington luminosity. Then

$$\eta c^2 \dot{M}_{\mathrm{acc}} \leq L_{\mathrm{Edd}} = \frac{4\pi G M c}{\kappa}. \tag{45}$$

Not all of this accretion goes to growing the black hole mass M, since an amount $L/\eta c^2$ of its rest-mass energy is lost to radiation. So the rate of black hole mas growth is

$$\dot{M} = (1 - \eta)\dot{M}_{\rm acc} \tag{46}$$

Rearranging we get

$$\dot{M} \leq \frac{1 - \eta}{\eta} \frac{M}{t_{\rm Edd}}, \tag{47}$$

where

$$t_{\rm Edd} = \frac{\kappa c}{4\pi G} = 4.5 \times 10^8 \ {\rm yr}. \tag{48}$$

Integrating through the inequality (47) we get

$$\frac{M}{M_0} \leq \exp\left[\left(\frac{1}{\eta} - 1\right)\frac{t}{t_{\rm Edd}}\right]. \tag{49}$$

At a lookback time $t = 10^9$ yr (redshift $z = 6$) we have $t/t_{\rm Edd} = 2.2$, and so

$$\frac{M}{M_0} \leq \exp\left[2.2\left(\frac{1}{\eta} - 1\right)\right]. \tag{50}$$

This innocent-looking expression is extraordinarily sensitive to the value of η and hence to the SMBH spin parameter a. For a maximally spinning hole ($a = 1$) we have $\eta = 0.42$ and

$$\frac{M}{M_0} \leq e^3 = 21, \tag{51}$$

so only a little growth, whereas for the usual 'default' efficiency $\eta = 0.1$ ($a = 0.85$) this limit changes to

$$\frac{M}{M_0} \leq e^{19.8} = 4 \times 10^8 \tag{52}$$

while for $a = 0$, ($\eta = 0.057$) we get a prodigious

$$\frac{M}{M_0} \leq e^{36} = 8 \times 10^{15}. \tag{53}$$

So the minimum initial masses at $z >> 1$ needed to grow to the observed $M > 5 \times 10^9 {\rm M}_\odot$ at $z = 6$ are

$$M_0 > 2 \times 10^8 {\rm M}_\odot \ \text{for maximal spin a} = 1, \tag{54}$$

but only

$$M_0 > 1.25 {\rm M}_\odot \ \text{for a} < 0.85 \tag{55}$$

with a limit well below a stellar mass for the non-spinning case $a = 0$. In other words, *unless black holes consistently spin rapidly* ($a > 0.9$) *and always accrete prograde, they can grow to the masses seen at* $z = 6$ *by accretion from initial* **stellar** *masses without violating the Eddington limit* (Figs. 21 and 22).

One might guess that if accretion is chaotic, the fact that there is no obvious reason for it to be prograde rather than retrograde wrt the SMBH spin would suggest that these spins remain low, as retrograde episodes could cancel the spinup effect of prograde accretion. The resulting low radiative efficiency η then means that growth to the quasar masses deduced at redshift $z = 6$ from initial stellar masses is perfectly possible, given an adequate mass supply. This guess is right.

A detailed calculation [36] considers repeated accretion disc events randomly inclined to the SMBH spin. The discs are limited by self-gravity (see Sect. 4.2), which shows that the condition $J_d < 2J_h$ (cf Eqs. 42, 43) is usually easily satisfied, allowing both co- and counter-alignment depending on the initial disc orientation. The results show that the guess that chaotic accretion keeps the SMBH spin low is correct, allowing rapid SMBH mass growth (see Fig. 23).

Some features of this figure are worth noting. First, if in the course of a merger with another galaxy the SMBH coalesces with another hole, its mass grows significantly in a very short time, and its spin may change rapidly too. In terms of the figure the SMBH jumps discontinuously off the spin evolution curve. But the curve is an attractor, so SMBH return to it once they have roughly doubled their masses by accretion. However for the very largest SMBH, typically found in the nuclei of giant

Fig. 21 Gas falling towards the central supermassive black hole can be both prograde and retrograde wrt the black hole spin, with equal probabilities

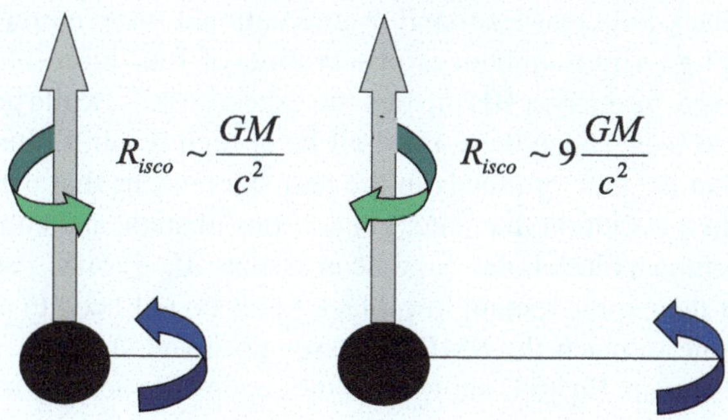

Fig. 22 For a maximally-rotating black hole, the ISCO has very different properties for prograde and retrograde orbits. Accreting from a retrograde orbit has a bigger effect since the last stable orbit has a larger lever arm than prograde one

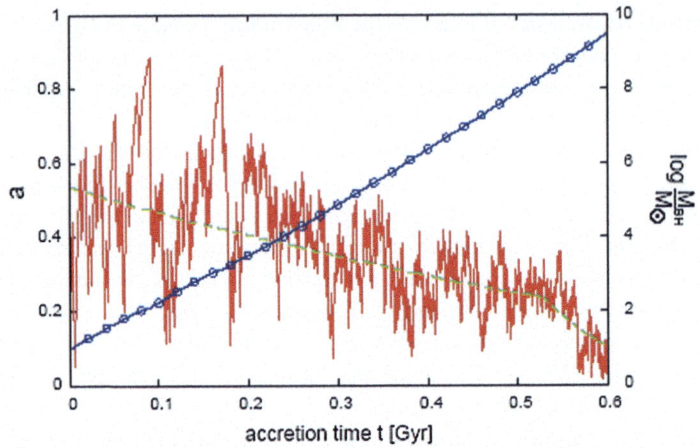

King, Pringle & Hofmann, MNRAS 2008

Fig. 23 The evolution of the mass and spin of a chaotically-accreting black hole

ellipticals, there is too little gas to fuel mass-doubling after a merger. So a suitably-arranged merger can give some SMBH in giant ellipticals high spin rates $a \sim 1$ which they will retain, whereas in most other cases SMBH revert to the spin evolution curve and so should spin slowly. Note that chaotic accretion does not quite reduce the spin to zero, but to a small value determined statistically by the property that successive accretion events can cause the spin to oscillate between small prograde and retrograde values.

One might worry that the calculation above implicitly assumed that accretion at the Eddington rate was continuous for the whole of the lookback time 10^9 yr since the Big Bang, and also took no account of the time needed to grow the initial stellar mass black hole. These assumptions are actually not so unreasonable, as one might expect the most extreme SMBH masses to correspond to the most extreme accretion regimes, but in any case one can easily relax these assumptions. The lifetime of a star massive enough to end its evolution as a stellar-mass black hole is $<10^8$ yr, significantly shorter than the lookback time 10^9 yr at $z = 6$. The recent compilation of black-hole masses found in gravitational-wave mergers (https://physicsworld.com/a/ligo-virgo-announces-observation-of-four-more-black-hole-mergers/) shows that even *pre*-merger BH masses can exceed $50\mathrm{M}_\odot$, while post-merger masses up to $80\mathrm{M}_\odot$ are seen. Since these must all be at high redshift, this confirms the long-held view that the lack of metals in the first stars means that they lose much less mass during their evolution than later generations of stars, and consequently leave much heavier remnant black holes. So we can assume $M_0 \gtrsim 50\mathrm{M}_\odot$ above. Then it is easy to see that if their spins remain low, black holes do not need to accrete for the entire lookback time to reach the SMBH masses observed at $z = 6$: assuming $a = 0$ for a simple estimate, Eq. (53) shows that the required timescale is less than one-half of this.

6.2 *Massive Seeds?*

If instead of this chaotic accretion picture we had assumed that the holes always accreted prograde and so always had high spins, significant mass growth from accretion would have been impossible. Then maintaining the assumption that mass growth at super-Eddington rates is impossible would force us to conclude that their masses must already have been very large at high redshift (larger than most SMBH at $z = 0$!). This apparent need for these massive seed black holes would be a direct consequence of the high radiative efficiency of rapidly-spinning black holes, meaning that the Eddington luminosity corresponds to a rather low accretion rate. In fact this view, that massive initial seeds were required to explain the SMBH masses at redshift 6 and beyond, prevailed for some time. This was despite a general recognition that matter accreting at large distances from the SMBH could not ‘know’ that it had to accrete prograde and spin the hole up.

The reason was the early erroneous treatment of disc-spin alignment we noted in Sect. 5.1 above. For a decade this treatment was unquestioned, with the consequent deduction that retrograde discs were impossible: after a very short time, *any* form of accretion on to an SMBH would become prograde. It is easy to see that adding disc matter with the specific angular momentum of the prograde ISCO spins the hole up to $a \simeq 1$ from any initial value once the mass of the hole has increased by a factor $\sim$2. So in this erroneous picture, all SMBH had maximal spin $a = 1$ almost all the time. As we have seen in Eq. (51) this meant that they could not have grown their masses by factors more than $\sim$20 by redshift 6. This erroneous result forced astronomers at the time to consider how massive seeds could form, and there is still work in this area. Of course the fact that growth from initial stellar masses is *possible* does not necessarily mean that there are no massive initial seeds, only that there is no pressing need for them.

6.3 *How Big Can a Black Hole Grow?*

Given a sufficient mass supply, it appears that SMBH can growth quite rapidly to very large masses—several are known with masses $\gtrsim 10^{10} \mathrm{M}_\odot$. This appears to be the limit on the mass of an SMBH which can undergo luminous accretion, i.e. gain mass from an accretion disc and radiate, making it directly observable as an AGN. We can show that there is a good reason for this.

For thin discs around SMBH the disc scaleheight obeys $H/R \sim 10^{-3}$ [9, 36]. Then using all the disc equations gives the radius where the disc is cut off by self-gravity as

$$R_{\mathrm{sg}} = 3 \times 10^{16} \alpha_{0.1}^{14/27} \eta_{0.1}^{8/27} (L/L_{\mathrm{Edd}})^{-8/27} M_8^{1/27} \ \mathrm{cm}. \tag{56}$$

Here L is the accretion luminosity and $L_{\rm Edd}$ the Eddington luminosity [9, 35]. Further, $\alpha = 0.1\alpha_{0.1}$, $\eta = 0.1\eta_{0.1}$ are the standard viscosity parameter, and the accretion efficiency respectively, and $M_8 = M/10^8{\rm M}_{\odot}$.

We know that the outer radius of any SMBH accretion disc cannot exceed $R_{\rm sg}$, which from (56) is effectively independent of the SMBH mass. But any disc must be at least as large as the ISCO, and its size scales directly with the SMBH mass M, as

$$R_{\rm ISCO} = f(a)\frac{GM}{c^2} = 7.7 \times 10^{13} M_8 f_5\,{\rm cm}. \qquad (57)$$

Here $f(a)$ is a dimensionless function of the spin parameter a, as we have seen in Sect. 2, and I write $f(a) = 5f_5(a)$, so that $f_5 \simeq 1$ describes prograde accretion at moderate SMBH spin rates $a \simeq 0.6$. If $R_{\rm ISCO} > R_{\rm sg}$, disc accretion is likely to be very weak or entirely absent. If the SMBH accretes any matter at all, it must be self-gravitating, and so swallowed whole, without dissipating through viscosity and radiating as a disc. An SMBH might in principle grow its mass in this way, but is very unlikely to appear as a bright disc-accreting object, i.e. a quasar or AGN.

Combining (56) and (57) we see that disc formation is impossible (because the ISCO radius exceeds the self-gravity radius) for SMBH masses larger than

$$M_{\rm max} = 5 \times 10^{10}{\rm M}_{\odot}{\alpha_{0.1}}^{7/13}{\eta_{0.1}}^{4/13}(L/L_{\rm Edd})^{-4/13}f_5^{-27/26} \qquad (58)$$

King [24]. This is an upper limit to the mass of the SMBH in any quasar or AGN, since these systems must have accretion discs. SMBH can in principle still grow above this limit, for example by merging with another SMBH, or swallowing stars whole, but cannot appear as luminous galactic nuclei.

Figure 24 compares the limit (58) with observations. All the masses measured for accreting SMBH lie below the $M_{\rm max}(a)$ curve for all values of a, with two exceptions which slightly restrict the SMBH spin. For 0014 + 813 [18] estimate $M \simeq 4 \times 10^{10}{\rm M}_{\odot}$, and for H1821+643 Walker et al. [73] find $M \simeq 3 \times 10^{10}{\rm M}_{\odot}$. The first system obeys the limit (58) provided that accretion is prograde and $a > 0.2$. H1821+643 similarly obeys the limiting mass provided that $a > -0.1$—retrograde accretion on to this hole with $|a| > 0.1$ is ruled out, but prograde accretion is possible for any spin parameter $a > 0$,

Assuming that the maximum observable luminosity of an AGN is the Eddington limit, the mass limit (58) implies a luminosity limit [27]

$$L_{\rm max} = 6.5 \times 10^{48}\alpha_{0.1}^{7/13}\eta_{0.1}^{4/13}f_5^{-27/26}\,{\rm erg\,s^{-1}} \qquad (59)$$

This prediction agrees very well with the highest QSO bolometric luminosity so far found, which appears to be in the recent WISE survey of hot, dust-obscured galaxies ('Hot DOGs': Assef et al. [2], Fig. 4). (The SEDs of Hot DOGs are generally dominated by a luminous obscured AGN.)

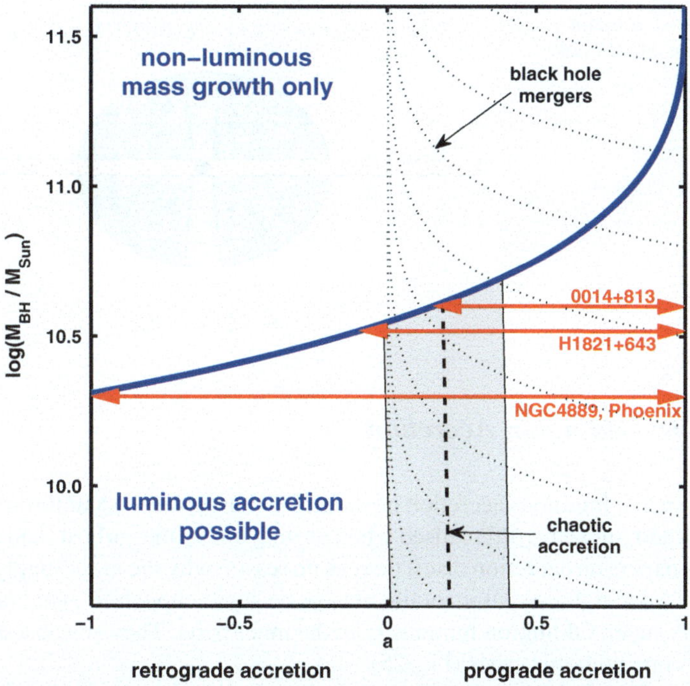

Fig. 24 The maximum mass for a black hole to have a luminous accretion disc

6.4 *AGN Variability*

The concept of chaotic accretion throws up another fairly basic consideration. We noted that if accretion involves individual disc episodes, the maximum mass that can be accreted each time is limited by self-gravity to $M_d \lesssim (H/R)M$, where M is the SMBH mass, and $H/R \sim 10^{-3}$ is the disc aspect ratio. Assuming that the discs frequently achieve masses of this order, and that the net result of each episode is mass accretion at rates $\dot{M}$ close to the Eddington value, this implies a characteristic timescale of AGN variability:

$$t_{\rm var} \sim \frac{M_d}{\dot{M}} \sim \frac{HM}{R\dot{M}} \sim 10^5\,{\rm yr} \tag{60}$$

King and Pringle [33], King and Nixon [26]. There is observational evidence that AGN vary (or flicker) on this timescale [64]—it appears to be the characteristic time for an AGN to light up through accretion after being quiescent.

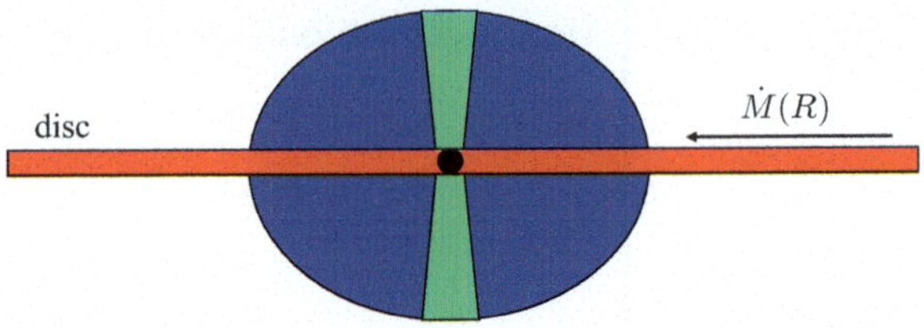

Fig. 25 Super-Eddington accretion on to a black hole

6.5 *Super-Eddington Accretion*

We have seen that luminous accretion on to SMBH must always occur through a disc. As Shakura and Sunyaev [66] realised when writing one of the earliest (and certainly most cited) papers on accretion discs, there is no reason why the mass supply rate $\dot M_0$ to the disc far from the accretor should always be small enough to avoid producing a potentially super-Eddington luminosity in the inner parts. They suggested how the disc might cope with this crisis (Fig. 25).

Clearly at disc radii $R > R_{\rm sph}$ with

$$\frac{GM\dot M_0}{R_{\rm sph}} = L_{\rm Edd}(M) = \frac{4\pi GMc}{\kappa} \tag{61}$$

there is no problem, as the local energy release is sub-Eddington, and radiation pressure is not able to overcome gravity. But within this radius this situation is reversed. Shakura and Sunyaev suggested that a possible outcome is that the excess radiation pressure could expel enough of the accreting matter at each radius R to keep

$$\frac{GM\dot M(R)}{R} = L_{\rm Edd}, \tag{62}$$

where $\dot M$ is now explicitly a function of R. Dividing these two equations shows that this requires

$$\frac{\dot M(R)}{\dot M_0} = \frac{R}{R_{\rm sph}} \tag{63}$$

with

$$\dot M_{\rm Edd} = \frac{L_{\rm Edd}}{\eta c^2} = \frac{4\pi GM}{\eta\kappa c} \tag{64}$$

The scaling radius

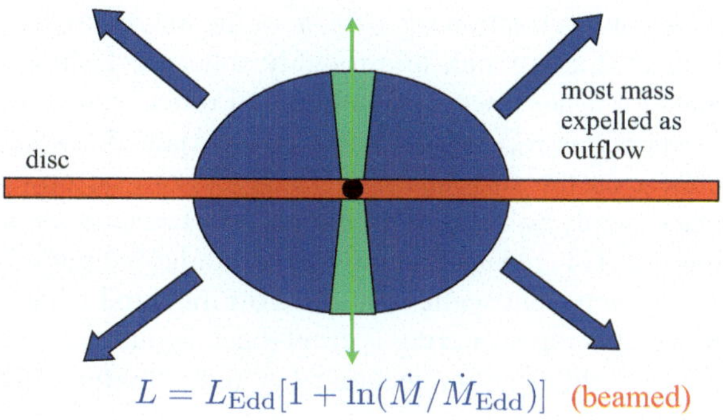

Fig. 26 Disc winds and radiation cones for super-Eddington accretion

$$R_{\rm sph} = \frac{GM\dot{M}_0}{L_{\rm Edd}} = \frac{\dot{M}_0\kappa}{4\pi c} \tag{65}$$

is often called the spherization or trapping radius, as the disc inflow is no longer thin but tending to become more spherical and marginally optically thick to electron scattering, so that photons cannot escape completely freely. It is now easy to show that the total accretion luminosity from the disc is

$$L \simeq L_{\rm Edd}[1 + \ln \dot{m}] \tag{66}$$

where $\dot{m} = \dot{M}_0/\dot{M}_{\rm Edd}$, for $\dot{m} > 1$—cf Fig. 26.

From Eqs. (63), (66) we conclude that in this picture a super-Eddington mass supply rate leads to mass gain by the accretor at just the Eddington rate, and raises the total accretion luminosity above Eddington by only a logarithmic factor. But two other effects are potentially very significant.

First, the emitted radiation pattern: from (63) the outflow is greatest at the point $R_{\rm sph}$, and is less at smaller radii, where most of the accretion luminosity is released. The result is that the radiation pattern has two sharply different components. A luminosity $\sim L_{\rm Edd}$ is emitted roughly isotropically from $R_{\rm sph}$, and all the rest of the total (66) is emitted from inside this radius in directions away from the disc plane. For $\dot{m} >> 1$ this component is strongly beamed along the disc axis with solid angle $4\pi b$, where

$$b \simeq \frac{73}{\dot{m}^2} \tag{67}$$

King [30]. The origin of this dependence is that while conditions near the accretor (where most of the accretion luminosity is released) are independent of $\dot{m}$, the outflow from radii $\sim R_{sph}$ remains optically thick to electron scattering up to a height above the disc plane which increases with $\dot{m}$, forcing radiation back towards the disc axis. Independently, one can derive the formula (67) empirically from the behaviour of soft X-ray components in ULX spectra, which often show a correlation $L \sim T^{-4}$ between blackbody luminosity and temperature [30].

Second, a fraction $(\dot{m}-1)/\dot{m}$ of the mass initially falling in towards the accretor is ejected as a wind, in a roughly spherical fashion, mostly from $R \sim R_{\rm sph}$. This matter reaches a terminal velocity of order the escape velocity $v = (GM/R_{\rm sph})^{1/2}$ of the radius $R_{\rm sph}$ where most of it is expelled, so the outflow can have very strong effects on the surroundings. These depend on how super-Eddington the original mass supply is. For $\dot{m} \sim 1$ the outflow has electron-scattering optical depth $\sim$1 to infinity. This means that most of the radially-emitted photons in the near-isotropic $L_{\rm Edd}$ component scatter just once off the wind material before escaping to infinity. Since electron scattering is front–back symmetric, a stream of photons on average transfers *all* its net momentum L/c to the outflow (where L is the luminosity), and we can write

$$\dot{M}v = \frac{L_{\rm Edd}}{c} \tag{68}$$

King [21]. Using $L_{\rm Edd} = \eta c^2 \dot{M}_{\rm Edd}$ we can rewrite this equation as

$$v \sim \frac{\eta}{\dot{m}} c \sim 0.1c, \tag{69}$$

since the accretion efficiency $\eta \sim 0.1$, and $\dot{m} \sim 1$, so we expect outflows from such systems with speeds $v \sim 0.1c$ [37]. It is important to note that the mechanical luminosity of a wind like this is significantly below $L_{\rm Edd}$:

$$L_{\rm mech} = \frac{1}{2}\dot{M}v^2 = \frac{1}{2}\dot{M}v.v = \frac{\eta}{2\dot{m}}L_{\rm Edd} \sim 0.05 L_{\rm Edd}. \tag{70}$$

If instead $\dot{m} >> 1$ the outflow must be very optically thick to electron scattering, making the radiation field mix strongly with it. Then the unbeamed luminosity $\sim L_{\rm Edd}$ shares its energy with the outflow, i.e.

$$L_{\rm mech} = \frac{1}{2}\dot{M}v^2 \simeq L_{\rm Edd} \tag{71}$$

Then we find

$$v \sim \left(\frac{2\eta}{\dot{m}}\right)^{1/2} c, \tag{72}$$

in place of Eq. (69) [25]. This also gives velocities $v \sim 0.1c$, but now with much more energy than (70).

All of these properties of super-Eddington accretion apply independently of the accretor mass M, so are in principle equally valid for stellar-mass accreting binary systems and for SMBH accretion. Stellar-mass binaries actually give very striking confirmation of these ideas. First, the basic result that the accretor gains mass only at the Eddington rate despite being fed at considerably higher rates is strongly confirmed by observations of X-ray binaries, particularly the unusual system Cygnus X-2 [38]. The accretor here is actually a neutron star rather than a stellar-mass black hole, but has a very similar accretion efficiency, so for most purposes the details of the accretion process are effectively identical. Stellar-evolutionary considerations show that the accretor in Cyg X-2 must have been fed a total mass $\sim 3\mathrm{M}_\odot$ from the companion star (more than twice its current measured mass!) over a timescale $\sim 10^6$ yr at rates $\gtrsim 10^2$ times its Eddington rate ($\sim 10^{-8}\mathrm{M}_\odot\,\mathrm{yr}^{-1}$). But it is clear that it gained very little of this mass, compatible with the expected $\sim 10^{-2}\mathrm{M}_\odot$—a direct illustration that at least in this case, a super-Eddington accretion mass supply only grows the accretor mass at the Eddington rate at most. As a second independent confirmation, the picture above implies that stellar-mass accretors with $\dot{m} >> 1$ must have apparent luminosities $>>L_{\mathrm{Edd}}$ if viewed along lines of sight close to the disc axis. This is by now widely accepted as the explanation of the ultraluminous X-ray sources (ULXs). These are found in external galaxies, and have apparent (assumed spherical) luminosities $\sim 10^{39}$–10^{41} erg s^{-1}, but are not located in the galaxy centres, and so cannot be supermassive black holes. In further agreement with the picture described above, all well-studied ULXs show evidence of strong winds with speeds compatible with (72).

In contrast, no SMBH analogues of ULXs are known so far, and we shall see instead that even the most luminous AGN are likely to have modest values $\dot{m} \sim 1$, at least in the low-redshift Universe.

6.6 *Super-Eddington Mass Growth?*

The discussion above led to the conclusion that feeding a black hole at a potentially super-Eddington rate can result in an outflow reducing the mass inflow rate to just the Eddington value at the accretor. It is important to ask if this is the only possibility, or whether in some parameter regimes an accreting black hole can grow its mass more rapidly than at the Eddington rate. One motive for much of the work on this possibility has been the perceived difficulty in growing large SMBH masses at redshifts $z \sim 6$. As we have seen, this hurdle may well be removed if SMBH accretion is sufficiently chaotic to keep the SMBH spin low. But I will point out later another possible reason why super-Eddington mass growth might be desirable in understanding how SMBHs get their observed masses.

The most obviously promising line for explaining super-Eddington mass growth comes from the simple argument that at sufficiently high mass supply rates, it seems plausible that the accretion flow is very optically thick to the radiation produced by its own viscous dissipation. This radiation is then trapped within the flow, and not easily available to drive away super-Eddington components of the flow as envisaged by Shakura and Sunyaev [66]. Instead it may be accreted and add to the black hole mass. Simple arguments (e.g. Ohsuga et al. [53]) suggest that photons generated near the midplane of discs with accretion rates $\dot{M} \gtrsim \dot{M}_{\mathrm{Edd}}$ are trapped and dragged into the black hole for disc radii

$$R < R_{\mathrm{trap}} \lesssim \dot{m} R_g \tag{73}$$

where $\dot{m} = \dot{M}/\dot{M}_{\rm Edd}$ is the Eddington accretion factor, $R_g = GM/c^2$ the black hole gravitational radius. The analytic 'slim disc' model [1] estimates the effect of trapping in reducing the luminous output of a disc. Numerical treatments (e.g. Ohsuga et al. [53]) find larger effects, which become very significant for $\dot{m} \gtrsim 10^2$ (note that the alternative definition $\dot{m} = \dot{M}c^2/L_{\rm Edd}$ used by that paper is larger than $\dot{m}$ defined above by a factor $1/\eta \sim 10$). Currently there is no clear consensus as to how large trapping effects can be, with differing numerical techniques finding a range of values. A further problem is the existence of at least two possible types of steady accretion flows (Shakura–Sunyaev, with ejection, and subluminous inflow with trapped photons) for mass supply rates $> \dot{M}_{\rm Edd}$. The way to discover which (if any) solution applies under given conditions should really be solved by using time-dependent calculations, but these are probably impossible analytically and currently intractable numerically. It is simple to see a potential physical instability choosing between the two types of solutions, as increased photon trapping presumably reduces gas ejection, and vice versa. This may suggest that the ejection picture is adequate for Eddington factors $\dot{m}$ not too large compared with unity, perhaps giving way to trapping solutions at larger $\dot{m}$, but so far no coherent treatment on these lines has been possible. This is an area of active current research (see Mayer [47] for a review). As mentioned above, there is observational support (from stellar-mass accreting systems) for the Shakura–Sunyaev + ejection picture, while this is more difficult for trapping. Accordingly in the rest of this article I adopt the ejection picture for the modest Eddington ratios likely in most observed SMBH systems, but bear in mind that the more drastic effects of trapping might appear at larger $\dot{m}$. This approach turns out to be fairly successful.

7 Black Hole Winds

The simple ejection theory of the last section strongly suggests that AGN should have wide-angle winds with $v \simeq 0.1c$ carrying the Eddington momentum (cf Eq. 68). This opens up the possibility that these winds provide the mechanical feedback needed to explain the SMBH—galaxy scaling relations (17), (18) discussed in Sect. 3.

Winds like this have been observed from most local AGN because of P Cygni profiles seen in the X-ray, where the resonance lines (usually of hydrogen—or helium-like iron) show blueshifted absorptions corresponding to $v \sim 0.1c$. These are now known as UFOs (ultrafast outflows)—cf Fig. 27.

The X-ray line ratios also give the ionization parameter

$$\xi = \frac{L_i}{NR^2}, \tag{74}$$

where N is the electron number density at distance R from the ionizing source (here the AGN) and L_i the part of the source luminosity consisting of photons whose energies are high enough to ionize the atomic species producing the observed lines. Since ξ and L_i are observable, (74) gives the quantity NR^2, and in addition the velocity v of a wide-angle wind is directly observed from the blueshifted lines. Since we expect any wind to accelerate rapidly to a terminal velocity v of order the escape speed from the launch point, and thereafter remain at this speed, we can estimate the total mass outflow rate

$$\dot{M}_{\rm out} = 4\pi b R^2 N \mu m_H v \tag{75}$$

where $b \sim 1$ is the fractional solid angle of the wind, and $\mu m_H \sim 0.6 m_H$ is the mean mass per electron, with m_H the hydrogen atom mass. With typical values $\xi \sim 10^4\,{\rm erg\,cm\,s^{-1}}$ and $v \sim 0.1c$ we find

$$\dot{M}_{\rm out} \sim 1 - 10\,{\rm M_\odot\,yr^{-1}}. \tag{76}$$

This estimate assumes that the solid angle factor $b \sim 1$, as is strongly suggested from the fact the essentially all sufficiently well-studied AGN show UFOs [69, 70]. This result is suggestive: the observed outflow rates are comparable with the accretion rates needed to make SMBH of masses $10^7{\rm M_\odot} \lesssim {\rm M} \lesssim 10^9{\rm M_\odot}$ radiate close to their Eddington luminosities. As the winds have the velocities $v \sim 0.1c$ expected from the simple disc wind theory we worked out in the last section, it is likely that these winds are what carries the feedback linking galaxy structure to SMBH mass growth. In line with this we find $\dot{M}_{\rm out} v \sim L_{\rm Edd}/c$, as expected if SMBH accretion rates approximate Eddington, but not significantly greater, when feedback occurs. This restriction is natural, as there is a limit on how rapidly gas can fall into an SMBH in a quasispherical galaxy bulge of velocity dispersion σ. The total mass of gas and stars at any radius R is given by (15), and we can expect only a fraction $f_g \sim 0.1$ of it is gas. Even if all centrifugal support were instantly removed, it would take a free-fall time $t_{\rm ff} \sim R/\sigma$ for this gas to reach the SMBH. Using (15) we see that the maximum gas accretion rate from any self-gravitating system characterized by a velocity dispersion σ is the dynamical rate

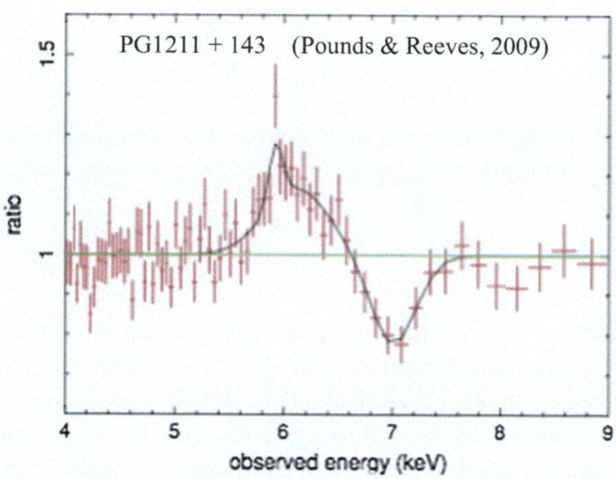

Fig. 27 Outflows driven by supermassive black holes: the 'P Cygni' profile of the line indicates a quasi-spherical ultra-fast outflow (UFO) with speed ~0.1c

$$\dot{M}_{\rm dyn} \sim \frac{f_g \sigma^2 R}{G} . \frac{\sigma}{R} \sim \frac{f_g \sigma^3}{G}. \tag{77}$$

We can compare this with the rate $\dot{M}_{\rm Edd}$ needed to give the Eddington luminosity for an SMBH if its mass is close to the $M-\sigma$ value. Using (4), (19) gives

$$\frac{\dot{M}_{\rm dyn}}{\dot{M}_{\rm Edd}} \sim \frac{35}{\sigma_{200}} \sim \frac{40}{M_8^{1/4}}. \tag{78}$$

Since $\dot{M}_{\rm dyn}$ is an upper limit derived for extreme assumptions, we see that for SMBH close to the $M-\sigma$ relation, strongly super-Eddington accretion is unlikely, in agreement with our conclusion above. (Of course it is conceivable that much more extreme conditions might occur when the black hole mass is much smaller than the $M-\sigma$ value.)

Given this result we see from (70) that the amount of SMBH binding energy potentially able to affect the host galaxy bulge is now reduced by a factor $\eta/2 \sim 1/20$ from the estimate (20), down to $E_{\rm BH} \sim 10^{60} M_8$ erg. But this is still uncomfortably large compared with the bulge binding energy $E_b \sim 8 \times 10^{58}$ erg (21). We shall see in Sect. 8 below how the efficiency of SMBH feedback is is further reduced, removing the danger that an SMBH could destroy its host galaxy.

7.1 Observability

X-ray spectra of UFOs give another observable quantity, the X-ray photoelectric absorption column N_H, which is given by the turndown in the continuum at low photon energy. This is a significant constraint on the wind, since from the mass conservation Eq. (75) we expect

$$N_H = \int_{R_{\rm in}}^{\infty} \frac{\dot{M}_{\rm out}}{4\pi R^2 b v m_H} {\rm d}R = \frac{L_{\rm Edd}}{4\pi b c v^2 m_H R_{\rm in}}, \tag{79}$$

where $R_{\rm in}$ is the inner radius of the flow, and we have used (68) at the last step. From the definition of $L_{\rm Edd}$ we find the electron scattering optical depth of the wind as

$$\tau = N_H \sigma_T \simeq \frac{GM}{b v^2 R_{\rm in}} \tag{80}$$

with $\sigma_T \simeq \kappa m_H$ the Thomson cross-section. So if the wind were continuous, so that $R_{\rm in}$ is the launch radius $\sim GM/bv^2$, it would be marginally optically thick (i.e. $\tau = 1$) to electron scattering. In practice, all UFOs are found to have N_H significantly smaller than the value $N_H \sim 1/\sigma_T \simeq 10^{24}\,{\rm cm}^{-2}$ for a continuous wind. This is reasonable, since it is hard to detect such Compton-thick AGN, but it means that we are probably unable to see those AGN where wind feedback on the host galaxy is at its most effective.

UFOs actually instead lie in the range $N_{22} \sim 0.3 - 30$, where $N_{22} = N_H/(10^{22}\,\mathrm{cm}^{-2})$. This implies that they are seen when their inner radii are larger than the launch radius $\sim GM/bv^2$. This must mean that UFO winds are *episodic*, and we only see them some time after launch (see Fig. 28). The wind is actually a sequence of roughly spherical shells rather than a continuous flow. Since the column density is dominated by the inner radius of each shell, we probably only detect the inner edge of the last shell to be launched. The time t_{off} since launch is given by setting $R_{\mathrm{in}} = vt_{\mathrm{off}}$. Using (80) we get

$$t_{\mathrm{off}} = \frac{GM}{bv^3 N_H \sigma_T} \simeq \frac{3M_7}{bv_{0.1}^3 N_{22}} \text{ months} \tag{81}$$

where $M_7 = M/10^7\mathrm{M}_\odot$ (the typical value for UFO sources is $M_7 \sim$ a few) and $v_{0.1} = v/0.1c$. We see that detecting UFOs with $N_H > 30$ would require an observation within a day of the launch of the most recent outflow episode. So this observed upper limit to N_H in UFOs probably reflects the relatively sparse observational coverage possible in X-rays, together with the fact that UFOs are Compton-thick for N_H only a factor 3 greater than this.

The lower limit $N_H \sim 0.3$ for UFO detections is probably the minimum required for significant optical depth in the blueshifted absorption lines of hydrogen- and helium-like iron (see King and Pounds [29]; Sect. 3.3). From (81) this means that current X-ray coverage is unlikely to detect UFOs launched more than a few months in the past, because the blueshifted iron absorption is too weak. Even observed UFOs should presumably slowly become unobservable as N_H drops, if followed for a few years. We will see below that a UFO wind has to travel out $\sim 10M_7$ pc before feeding back its energy and momentum in a collision with interstellar medium of the host galaxy, which at a speed $\sim 0.1c$ takes a time $t_{\mathrm{coll}} \sim 300M_7$ yr. A final restriction on observability is simply that the outflow may be too highly ionized to be detectable, and significant N_H is unmeasurable.

These considerations mean that trying to use observed UFOs to constrain the physical state of the AGN which launched them is not straightforward. Because UFOs are episodic and AGN are variable, it is perfectly possible to observe the

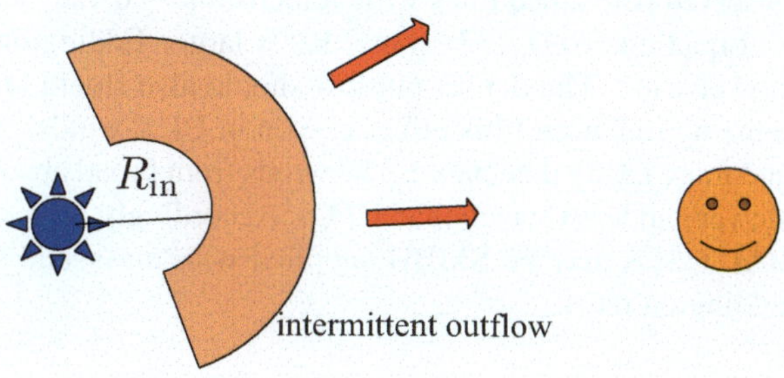

Fig. 28 Outflows from supermassive black holes: the column density measured in X-ray observations reveals their variable nature

AGN at a sub-Eddington luminosity even though we might expect a luminosity $\sim L_{\rm Edd}$ to be needed to launch its UFO. As an illustration, narrow-line Seyfert 1 galaxies show signs of super-Eddington phenomena but are usually observed to be sub-Eddington (cf NGC 4051; Denney et al. [12]). UFO episodes seem to be launched in brief phases of super-Eddington accretion, although the longer-term average accretion rate is sub-Eddington. So it is important to bear in mind how sparse the observational coverage of UFOs is. We can only see episodic UFO shells for tiny fractions $t_{\rm off}/t_{\rm coll} \sim 10^{-3} v_{0.1}^{-2} N_{22}^{-1}$ of the 300–3000 yr time it takes before they collide with the host interstellar medium. Most UFO episodes in any given source are undetected, and more AGN produce them than we are able to observe. Most seriously, since a continuous outflow is Compton-thick, we cannot directly observe the most active form of AGN feedback at all.

7.2 Wind Ionization and BAL QSOs

The X-ray spectrum of a UFO is determined by the ionization parameter ξ (Eq. 74). Eliminating the combination NR^2 between this equation and (75) and using (69) gives

$$\xi = 3 \times 10^4 \eta_{0.1}^2 \dot{m}^{-2} = 3 \times 10^4 v_{0.1}^2 l_2 \tag{82}$$

where I have parametrized $\eta = 0.1\eta_{0.1}$, $v = 0.1cv_{0.1}$ and $L_i = lL_{\rm Edd} = 10^{-2} l_2 L_{\rm Edd}$, since the ionizing luminosity is generally much smaller than the total AGN output. This equation relates the wind mass and speed to its ionization state. For a given quasar spectrum, the ionization state must be such that the threshold photon energy defining L_i and the corresponding ionization parameter ξ satisfy (82). We can see that (82) rules out low ionization: the low threshold would mean that most of the AGN radiation would ionize, making l_2 large. But (82) would then give a *high* ionization parameter, in contradiction. Depending on the AGN continuum spectrum (82) can have a range of solutions. If more than one is possible, initial conditions would determine which applies. One obvious solution is $\dot{m} \sim 1$, $l_2 \sim 1$, $\xi \sim 3 \times 10^4$, which gives the conditions one would expect to see if an observation caught a UFO very close to launch. In most UFO observations the AGN luminosity has dropped below $L_{\rm Edd}$ after launching a wind shell, and at say $0.3L_{\rm Edd}$ we get $\xi \sim 10^4$, sufficient to produce helium- and hydrogen-like iron. This agrees with the presence of the observed resonance lines with velocities $v \sim 0.1c$.

Equations (69), (72) show that a larger Eddington factor $\dot{m}$ should produce a slower wind. The denser outflow should also shield large parts of the outflow from ionizing radiation (this effect is seen in ULX winds), which should be less ionized and more easily detectable. The winds from broad absorption line (BAL) quasars are slower and less ionized than UFOs. Accordingly Zubovas and King [75] suggest that BAL QSOs may be SMBH supplied with mass–and so also ejecting it—at super-Eddington rates.

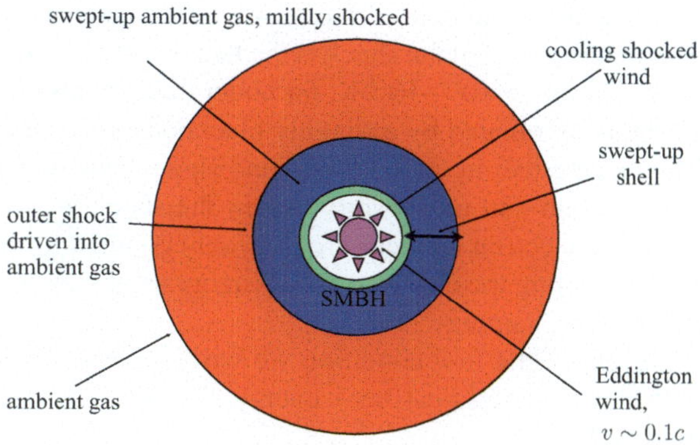

Fig. 29 The effect of black hole outflows on the surrounding gas

8 The Wind Shock and the $M - \sigma$ Relation

8.1 Momentum or Energy?

We have seen that SMBH accreting at mildly super-Eddington rates produce wide-angle winds with speeds $v \sim 0.1c$, and total (scalar) momentum

$$\dot{M}_{\rm out} v \simeq \frac{L_{\rm Edd}}{c} \tag{83}$$

comparable to the photon momentum of the AGN. Inevitably this wind must shock as it collides with the interstellar medium of the host bulge, converting a large fraction of the wind kinetic energy into thermal energy. The collision also drives a forward shock into the interstellar gas, so we get two shocks each side of the contact discontinuity where the wind and interstellar case meet, qualitatively very similar the the effect of a stellar wind on the interstellar medium (see Fig. 29).

The effect of the collision on the bulge gas is sharply different depending on whether the shocked wind gas radiates away its newly-increased thermal energy on a timescale much shorter than the postshock flow timescale or not. The first case, where radiation cools the postshock gas very efficiently, is often called an isothermal shock, as the temperature of the shocked gas rapidly returns to something like its preshock value. Its thermal pressure is then much smaller than its ram pressure, which has the same value (83) as before the shock. The cooled wind gas lies in a very thin layer and sweeps up the ambient interstellar gas ahead of it. This is sometimes called momentum-driven flow (but note that this same term is also used for a very different physical situation when radiation is incident on dust—see Sect. 7.4.2 of King and Pounds [29] for a discussion).

If instead radiative cooling is ineffective, the postshock thermal pressure is much stronger than the ram pressure, and does work against the ambient interstellar gas as it expands adiabatically. This case is called energy-driven (or adiabatic) flow.

We will see that both of these cases occur, and have very different effects. The process tending to cool the shock is the fact that the electrons of the shocked wind reach a temperature $\sim 10^{10}$–10^{11} K, far hotter than the photons of the AGN radiation field, which is dominated by a roughly blackbody component with temperature $\sim 10^7$ K. This is effective in cooling the wind shock provided that the density of photons is high enough—in other words, when the shock is close enough to the AGN—and when the electron temperature is closely coupled to that of the ions. This is needed so that cooling the electrons takes energy from the ions, which carry the vast bulk of the postshock energy.

To check the first condition we write the Compton cooling time of an electron with the expected postshock energy $E \simeq 9m_p v^2/16$ in a radiation field of energy density $U_{\rm rad}$ as

$$t_C = \frac{3m_e c}{8\pi\sigma_T U_{\rm rad}} \frac{m_e c^2}{E} \tag{84}$$

where m_e is the electron mass. We assume

$$U_{\rm rad} = \frac{L_{\rm Edd}}{4\pi R^2 c} \tag{85}$$

where R is the distance of the electron from the accreting SMBH. Substituting for $L_{\rm Edd}$ we get

$$t_C = \frac{2}{3}\frac{cR^2}{GM}\left(\frac{m_e}{m_p}\right)^2\left(\frac{c}{v}\right)^2 \simeq 10^7 R_{\rm kpc}^2 M_8 \,{\rm yr}. \tag{86}$$

To find out if we have momentum- or energy-driven flow we have to compare this with the shock travel time $\sim R/\dot{R}_s$. For momentum-driven flow we will find below that $\dot{R}_s \sim \sigma$, so comparing with (86), momentum-driven flow holds self-consistently for wind shocks within a radius

$$R_C \simeq 0.5 \frac{M_8}{\sigma_{200}} \,{\rm kpc} \tag{87}$$

which we call the cooling radius. Physically, the radiation field becomes too dilute to cool the electrons for $R > R_C$. As we noted above, this estimate assumes that the electrons and ions in the postshock flow are in equipartition, which can be questioned [16, 50] on the grounds that Coulomb collisions may be too infrequent. But this is not the only way of bringing about equipartition—other processes are very likely to do this.

In addition, there is observational evidence of the Compton continuum removing most of the shock energy (as required for momentum-driven flow) in the one source where we see direct evidence of a wind shock. Schematically the wind shock should have the structure shown in Fig. 30.

In particular, the gas velocity should decrease markedly at the shock, and if cooling is sufficient to bring about momentum-driven flow, the gas temperature should

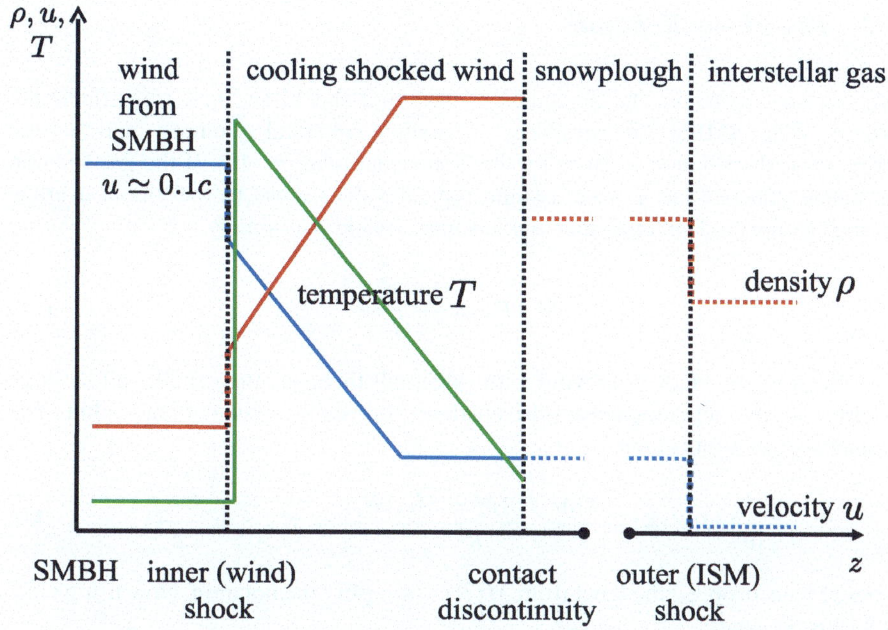

Fig. 30 Impact of a wind from an SMBH accreting at a super-Eddington rate on the interstellar gas of the host galaxy. Schematic view of the radial dependence of the gas density, velocity and temperature. At the inner shock, the gas temperature rises strongly while the wind density and velocity, respectively, increase (decrease) by factors of ∼4. Immediately outside this (adiabatic) shock, the strong Compton cooling effect of the quasar radiation severely reduces the temperature, and slows and compresses the wind gas still further. This cooling region is very narrow compared with the shock radius, and may be observable through the inverse Compton continuum and lower excitation emission lines. The shocked wind sweeps up the host ISM as a 'snowplough'. This is more extended than the cooling region, and itself drives an outer shock into the ambient ISM of the host. The vertical dashed lines denote the three discontinuities, inner shock, contact discontinuity and outer shock

decrease along with it. This is observed in NGC 4051 [57], where the ionization parameter is used as a proxy for the temperature. Observations by Pounds and Vaughan [59] further show a Compton hump in the low-energy X-ray spectrum which has the luminosity $\sim(\eta/2)L_{\rm bol}$ expected if it removes most of the postshock heat energy from the wind gas. Evidently this is the only system where we are lucky enough to observe directly the impact of the black hole wind on the host galaxy interstellar medium.

8.2 The $M-\sigma$ Relation

We can now work out the dynamics of the wind shock, and how this affects the growth of the SMBH. For simplicity we assume spherical symmetry, and that the bulge mass distribution is given by the isothermal relation (15). If the gas fraction (assumed constant) is f_g (we assume typical values close to the global average $f_g \simeq 0.16$ for the Universe) then the gas mass inside R is simply

$$M_g(R) = \frac{2 f_g \sigma^2 R}{G}. \tag{88}$$

The effect of the black hole wind is to sweep all this gas into a shell, and if shock cooling is efficient (momentum-driven flow) this shell is geometrically thin. The equation of motion is then

$$\frac{\mathrm{d}}{\mathrm{d}t}[M(R)\dot{R}] + \frac{GM(R)[M + M_{\rm tot}(R)]}{R^2} = \dot{M}_{\rm out} v = \frac{L_{\rm Edd}}{c}, \tag{89}$$

where I have dropped the label from $M(R)$ to simplify the notation. Now using (15), (88) this becomes

$$\frac{\mathrm{d}}{\mathrm{d}t}(R\dot{R}) + \frac{GM}{R} = -2\sigma^2\left[1 - \frac{M}{M_\sigma}\right] \tag{90}$$

where

$$M_\sigma = \frac{f_g \kappa}{\pi G^2}\sigma^4 \tag{91}$$

We can integrate this equation by multiplying through by $R\dot{R}$, to get

$$R^2\dot{R}^2 = -2GMR - 2\sigma^2\left[1 - \frac{M}{M_\sigma}\right]R^2 + \text{constant} \tag{92}$$

We see that if $M < M_\sigma$ there is no solution for large R, as the rhs becomes negative. Physically this happens because the small SMBH mass means that the wind thrust $L_{\rm Edd}/v$ is too small to lift the gas to large radius. If instead $M > M_\sigma$ we see that for large R, the velocity $\dot{R}$ tends to a constant value $\propto \sigma$, so the swept-up gas can be expelled to large radius. Intuitively it is suggestive that once the SMBH reaches the value (91), significant further mass growth is prevented because the gas that could fuel it is driven away. The first derivation on these lines was by King [21], who got just one half of the expression (91) by asking for what mass a momentum-driven outflow could achieve a radial speed $\dot{R} = \sigma$, comparable with the escape value.

Remarkably, with $f_g = 0.16$, $\kappa = 0.34$ we find

$$M_\sigma = 3 \times 10^8 \mathrm{M}_\odot \text{œ}_{200}^4 \mathrm{M}_\odot \tag{93}$$

in very good agreement with the observed value (e.g. Kormendy and Ho [41], Fig. 8), even though there is essentially no free parameter in the relation (91). Also, the problem of the mismatch between the SMBH and bulge binding energies we have noted throughout these notes, particularly at the end of the introduction to Sect. 7, has now disappeared. This is a direct result of the efficient shock cooling, which reduces the *effective* mechanical energy $\sim(\eta/2)L_{\rm Edd} \sim 0.05L_{\rm Edd}$ by a further factor $\sigma/c \sim 10^{-3}$.

The very good agreement of (93) with observation may seem surprising, given the extreme simplicity of the derivation. First, even if we have spherical symmetry, the real galaxy potential could be more complex than isothermal. But multiplying through by $M(R)\dot{R}$ for any form of $M(R)$ always produces a first integral with properties like (92), specifying a critical SMBH mass for a momentum-driven outflow to escape, so things remain at least qualitatively similar. Further, specifying different explicit choices of potential leads to relations which are almost indistinguishable from (91) [48]. Even for potentials and gas distributions which are not spherically symmetric, SMBH momentum feedback still exerts strongly radial forces on gas exposed to it. We have already seen (Sect. 6.6) that accretion on to the central SMBH may well be chaotic, so that over time all directions receive similar feedback. Further, we will see shortly that once the critical (M_σ) mass is reached, the spatial scales of the outflow increase dramatically. These considerations may explain why the simple spherical treatment given here produces a surprisingly precise agreement with the observed critical mass. We will also see later (Sect. 8.4, after Eq. (107)) why observations tend to give an exponent for σ which is slightly larger than 4.

8.3 Near the Black Hole

Although I have discussed a single momentum-driven outflow pushing away much of the bulge gas and picking out the critical SMBH mass $M = M_\sigma$, there must have been many abortive outflows long before this, which pushed gas away from the vicinity of the SMBH, but did not have sufficient momentum to lift it decisively, and so eventually fell back. Gas in the region surrounding the SMBH must have been repeatedly compressed in this way. This has two important results.

First, some of this compressed gas must have formed stars. The timescale for growing the SMBH to M_σ must be several Eddington times, so more than 10^8 yr, long enough for generations of massive stars to finish their evolution and return much of their gas, now enriched with metals through stellar nucleosynthesis, to the local ISM near the black hole. Since the same gas is repeatedly swept up and enriched in this way, this explains why gas near AGN frequently shows *higher* metallicities than solar.

Second, the repeated movements of large masses (comparable to the SMBH mass) has gravitational effects, and may be what flattens the stellar distribution near the black hole (from a cusp to a core). Pontzen and Governato [56] earlier showed that repeated supernovae work in a similar way, but the much larger masses involved in momentum-driven episodes suggests a larger effect.

It is worth making a final remark here. We have got a long way by adopting the black-hole winds picture, which is essentially just the Shakura–Sunyaev ejection solution applied to supermassive systems. But a possible difficulty is that if the galaxy bulge grew more rapidly than the black hole mass, we might be in the regime where Eddington accretion factor $\dot{m}$ is $>> 1$, so that now Eq. (72) applies in place of (69). This apparently small change could have large effects: the greatly increased feedback strength it implies leads to a significantly smaller M_σ value, which could self-consistently keep $\dot{m} >> 1$ [25]. This would act as a bottleneck to SMBH growth, and produce a much lower-mass $M-\sigma$ relation than observed, although measuring these SMBH masses would be difficult. A simple way of avoiding this bottleneck would appear if sufficiently hyper-Eddington accretion of this type caused a transition to accretion discs dominated by photon trapping in the inner parts, as we discussed in Sect. 6.6. The black-hole mass would then grow with little inhibition from feedback effects, reducing $\dot{m}$ to values ~ 1, and so allowing it to reach the $M-\sigma$ mass (93).

8.4 What Happens at $M = M_\sigma$?

The expression (91) is so close to the observed value of the SMBH mass over a range of σ that reaching this value must trigger a shutdown of mass growth. This cannot be instant, as there may be gas in the immediate surroundings which is too dense to be driven away by the outflow—an accretion disc is a possible example, but the mass here is limited by self-gravity to $\sim (H/R)M << M$ (cf Eq. 33). But simply pushing gas away with a momentum-driven outflow alone would require a considerable increase in mass beyond M_σ to prevent later fallback [48, 67]. So we need the arrival of the SMBH at the M_σ mass to trigger a more drastic removal of gas from the bulge.

A candidate for this arises quite naturally [22]. Once a momentum-driven outflow is pushed beyond the cooling radius R_C (87) the wind shock is no longer effectively cooled by the inverse Compton effect, as the AGN radiation field becomes too dilute. The wind can now use all its thermal energy to push the bulge gas away at high speed. So once M reaches the value M_σ the outflow geometry changes radically (see Fig. 31).

The shocked wind region is no longer a thin layer, as outside R_C its high postshock thermal pressure drives rapid expansion outward into the bulge gas. This shock-heated gas has sound speed $\sim 0.03c$, which rapidly equalizes the pressure within it as it expands. The thermal pressure changes the equation of motion from the form (89) to

$$\frac{\mathrm{d}}{\mathrm{d}t}[M(R)\dot{R}] + \frac{GM(R)[M_{\mathrm{tot}}(R)]}{R^2} = 4\pi R^2 P, \tag{94}$$

where I have dropped the black hole gravity contribution to the second term as its mass is negligible compared with $M_{\mathrm{tot}}(R)$ for $R >> R_C > R_{\mathrm{inf}}$, and the pressure P

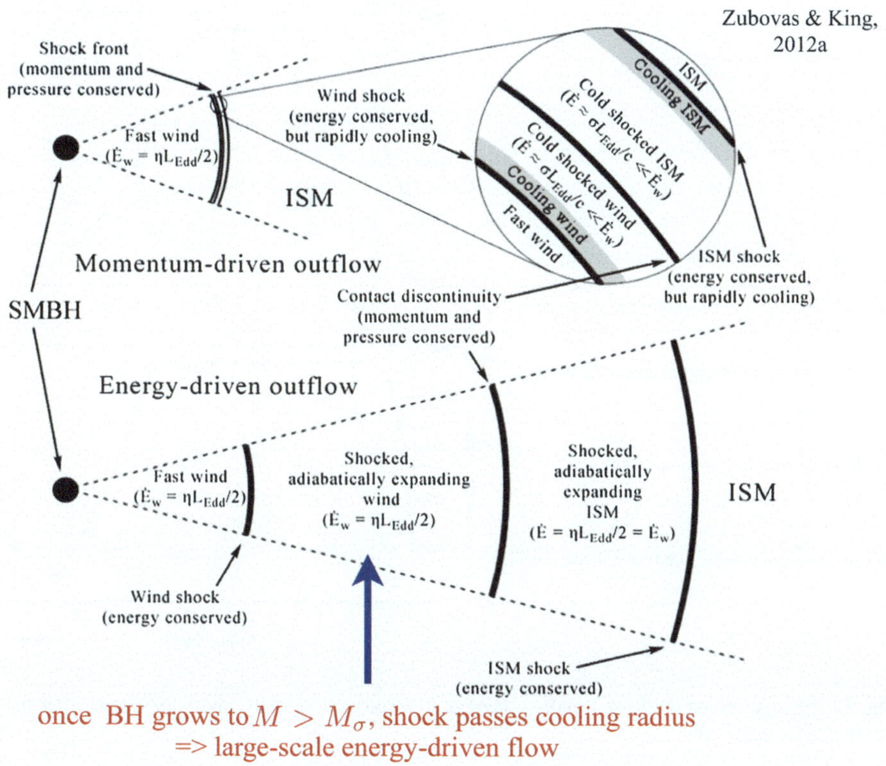

Fig. 31 The contrasting shock structure of momentum- and energy-driven black hole outflows. The blue arrow shows the transition from momentum- to energy-driven outflows as the M_σ mass is reached

is much larger than the ram pressure term $L_{\rm Edd}/c$ appearing in (89). We fix P by explicitly using the energy equation, which fixes the rate at which the shocked gas gains internal energy, reduced by $P\mathrm{d}V$ work on the ambient gas and against gravity:

$$\frac{\mathrm{d}}{\mathrm{d}t}\left[\frac{4\pi R^3}{3}.\frac{3}{2}P\right] = \frac{\eta}{2}L_{\rm Edd} - P\frac{\mathrm{d}}{\mathrm{d}t}\left[\frac{4\pi R^3}{3}\right] - 4f_g\frac{\sigma^4}{G}. \tag{95}$$

where I have assumed a specific heat ratio $\gamma = 5/3$. (We did not need the energy equation in the momentum-driven case. There it is equivalent to the condition that all energy except that associated with the ram pressure is lost to cooling.) Eliminating P between (94) and (95) gives

$$\frac{\eta}{2}lL_{\rm Edd} = \frac{2f_g\sigma^2}{G}\left\{\frac{1}{2}R^2\dddot{R} + 3R\dot{R}\ddot{R} + \frac{3}{2}\dot{R}^3\right\} + 10f_g\frac{\sigma^4}{G}\dot{R}, \tag{96}$$

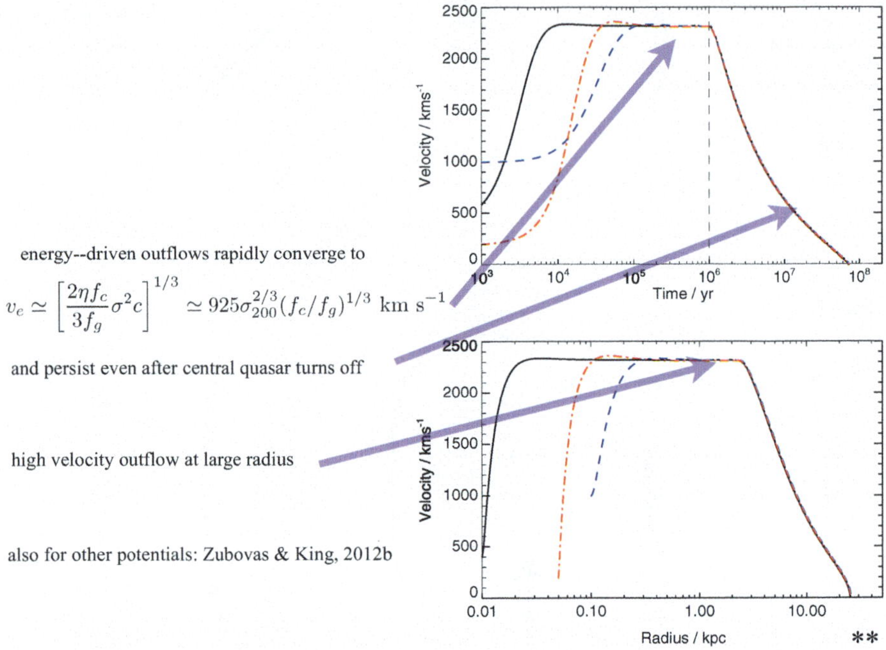

Fig. 32 Energy-driven outflows rapidly converge to a constant velocity, and persist long after the driving AGN turns off

where the quantity l allows for deviations of the AGN luminosity from $L_{\rm Edd}$ (see King [22]; King and Pounds [29] for details). This equation describes the motion of the wind/interstellar gas interface in the energy-driven case, replacing Eq. (89) in the momentum-driven case. We assume $M = M_\sigma$ in $L_{\rm Edd}$.

This formidable-looking equation has a simple solution $R = v_e t$, where

$$2\frac{\eta l c}{\sigma} = 3\frac{v_e^3}{\sigma^3} + 10\frac{v_e}{\sigma} \tag{97}$$

The second term on the rhs is negligible, as assuming the opposite, i.e. $v_e/\sigma << 1$, leads to a contradiction ($v_e \simeq 0.01c >> \sigma$). Then

$$v_e \simeq \left[\frac{2\eta l \sigma^2 c}{3}\right]^{1/3} \simeq 925 l^{1/3}\sigma_{200}^{2/3} \text{ km s}^{-1}. \tag{98}$$

This solution is an attractor: all solutions quickly converge to it, whatever their initial conditions, as we see from Fig. 32. If the AGN switches off ($l = 0$) with the contact discontinuity at radius R_0, we see from this figure that it continues to move out for a significant time. This agrees with the analytic solution of (98) for $l = 0$ [39]

$$\dot{R}^2 = 3\left(v_e^2 + \frac{10}{3}\sigma^2\right)\left(\frac{1}{x^2} - \frac{2}{3x^3}\right) - \frac{10}{3}\sigma^2 \tag{99}$$

where $x = R/R_0 \geq 1$. Because v_e depends on the AGN luminosity only as $v_e \sim l^{1/3}$ we see that variations or even disappearances of the AGN luminosity have almost no effect on the outflow once it is established. The flow persists because has a large reserve of thermal energy available for driving, and can last for a considerable time after the central AGN has switched off. This means that even outflows where no central AGN is observed may well be the result of one which happened in the past.

To work out the mass outflow rate we need to know how the outer shock overtakes the host ISM and entrains interstellar gas ahead of the expanding shocked wind. The ISM gas ahead of this forward shock is at rest, so this shock must overtake it at a speed giving a strong-shock velocity jump by a factor $(\gamma + 1)/(\gamma - 1)$ in the shock frame. (Here γ is the specific heat ratio—see for example Dyson andWilliams [13] for a derivation.) This gives the forward shock velocity as

$$v_{\rm out} = \frac{\gamma+1}{2}\dot{R} \simeq 1230\sigma_{200}^{2/3}\left(\frac{lf_c}{f_g}\right)^{1/3} \text{ km s}^{-1} \tag{100}$$

(with $\gamma = 5/3$ in the last step, and $f_c \simeq 0.16$ is the cosmological value of f_g). This gives a shock temperature $\sim 10^7$ K for the forward (ISM) shock (far lower than the $\sim 10^{10-11}$ K for the wind shock). The forward shock and the contact discontinuity were very close when energy-driven flow started (see Fig. 31) so the outer shock is at

$$R_{\rm out}(t) = \frac{\gamma+1}{2}R(t) = \frac{\gamma+1}{2}v_e t. \tag{101}$$

Then the mass outflow rate is

$$\dot{M}_{\rm out} = \frac{{\rm d}M(R_{\rm out})}{{\rm d}t} = \frac{(\gamma+1)f_g\sigma^2}{G}\dot{R}. \tag{102}$$

The black hole wind, with $M = M_\sigma$, has outflow rate

$$\dot{M}_{\rm w} \equiv \dot{m}\dot{M}_{\rm Edd} = \frac{4 f_c \dot{m}\sigma^4}{\eta c G}, \tag{103}$$

much smaller than the mass outflow rate $\dot{M}_{\rm out}$ from the shocked ISM it drives. The ratio of these two mass rates defines the mass-loading factor as

$$f_{\rm L} \equiv \frac{\dot{M}_{\rm out}}{\dot{M}_{\rm w}} = \frac{\eta(\gamma+1)}{4\dot{m}}\frac{f_g}{f_c}\frac{\dot{R}c}{\sigma^2}. \tag{104}$$

Then

$$\dot{M}_{\rm out} = f_{\rm L}\dot{M}_{\rm w} = \frac{\eta(\gamma+1)}{4}\frac{f_g}{f_c}\frac{\dot{R}c}{\sigma^2}\dot{M}_{\rm Edd}. \tag{105}$$

For $l \sim 1$ we have $\dot{R} = v_e$, and so from (98)

$$f_{\rm L} = \left(\frac{2\eta c}{3\sigma}\right)^{4/3}\left(\frac{f_g}{f_c}\right)^{2/3}\frac{l^{1/3}}{\dot{m}} \simeq 460\sigma_{200}^{-2/3}\frac{l^{1/3}}{\dot{m}}, \tag{106}$$

and

$$\dot{M}_{\rm out} \simeq 4060\sigma_{200}^{8/3}l^{1/3}\ {\rm M_\odot\,yr^{-1}} \tag{107}$$

for typical values $f_{\rm g} = f_{\rm c}$ and $\gamma = 5/3$. Equation (17) gives the total gas mass in the bulge as $M_g \sim 10^3 f_g M_\sigma$, so evidently if the outflow persists for a time $t_{\rm clear} \sim M_g/\dot{M}_{\rm out} \sim 1 \times 10^7\sigma_{200}^{2/3}l^{-1/3}$ yr it could potentially remove a large fraction of this gas. The time this actually takes depends on both the type and of the galaxy and its environment (field or cluster) and gives three parallel $M{-}\sigma$ relations with slightly different normalizations (see Zubovas and King [77]). Fitting a single relation $M \propto \sigma^\alpha$ to a sample containing all three types of galaxy tends to make the exponent slightly larger than the value $\alpha \simeq 4$ which holds for each separately.

Equations (98) or (99) specify the motion of the interface between the shocked wind and swept-up interstellar gas (see Fig. 31). This contact discontinuity is strongly Rayleigh-Taylor (RT) unstable: the shocked wind gas has greatly expanded, and so has much lower mass density than the interstellar gas it sweeps up, meaning we have heavy fluid higher in the gravitational potential than a light one. The RT instability drives strong overturning motions on all scales, making it hard to handle numerically, so we should be cautious about interpreting the predicted behaviour of the outflow. But the instability does explain why the high-speed ($\sim$1000 km s^{-1}) outflows we predict here usually have most of the outflowing gas in molecular form, which is what is observed (see below). A preliminary analysis Zubovas and King [78] suggested that most of the swept-up interstellar gas was likely to have a multiphase structure. Richings and Faucher-Giguére [61, 62] show that a full cosmochemical treatment indeed predicts that most of it cools all the way from the ISM shock temperature $\sim 10^7$ K to much lower temperatures. The swept-up gas is largely in molecular form, despite its entrainment in an outflow with the forward shock speed $\sim$1000 km s^{-1}.

We can ask where the black hole wind kinetic energy goes. From (100), (107) we get

$$\frac{1}{2}\dot{M}_{\rm w}v^2 \simeq \frac{1}{2}\dot{M}_{\rm out}v_{\rm out}^2. \tag{108}$$

As expected for energy driving the bulk of the wind kinetic energy ends up as mechanical energy of the outflow. Continuity of pressure and velocity across the contact discontinuity show that the shocked wind retains 1/3 of the total incident wind kinetic energy $\dot{M}_{\rm w}v^2/2$, while 2/3 goes to the swept-up ISM gas outflow.

As usual with mass-loading ($f_L > 1$), the scalar momentum rate of the swept-up gas is greater than of the wind driving it, which is already Eddington. From Eq. (108) we get

$$\frac{\dot{P}_{\rm w}^2}{2\dot{M}_{\rm w}} \simeq \frac{\dot{P}_{\rm out}^2}{2\dot{M}_{\rm out}}, \tag{109}$$

where the $\dot{P}$'s are the momentum fluxes. With $\dot{P}_{\rm w} = L_{\rm Edd}/c$, we find

$$\dot{P}_{\rm out} = \dot{P}_{\rm w}\left(\frac{\dot{M}_{\rm out}}{\dot{M}_{\rm w}}\right)^{1/2} = \frac{L_{\rm Edd}}{c} f_{\rm L}^{1/2} \simeq 20\frac{L_{\rm Edd}}{c}\sigma_{200}^{-1/3} l^{1/6}. \tag{110}$$

This agrees with observations: galaxy-wide molecular outflows are found in many galaxies, and have $\dot{M}_{\rm out} v_{\rm out} > L_{\rm Edd}/c$, and momentum rates $20L/c$ are common [8]. Cosmological simulations suggest that galaxies are still accreting mass at large scales [10], so these high momentum rates are important in preventing further mass growth.

Infrared observations give many examples of molecular outflows with speeds and mass rates similar to (100) and (107)—(see e.g. Cicone et al. [8], and the discussion in King and Pounds [29]), strongly suggesting that AGN feedback is the driver, as these mass and energy rates are generally too large to be driven by star formation.

We conclude that SMBH which have reached the M_σ mass can sweep their host galaxy bulges clear of gas. If this is true, we would expect the molecular outflows to have the mechanical luminosity (70):

$$L_{\rm mech} \sim \frac{\eta}{2} L_{\rm Edd} \simeq 0.05L, \tag{111}$$

where $L = lL_{\rm Edd}$ is the observed AGN luminosity. Cicone et al. [8], Fig. 12 do largely confirm the relation (111), but with exceptions. Further, there is a recent example [71] of a UFO wind simultaneously observed in the centre of a galaxy showing a large-scale outflow with the just the 'right' mechanical luminosity (111). But we should remember that while the molecular outflows are long-lived, the luminosity L of the central AGN is highly variable. We have already noted that outflows are seen with no AGN currently observable, but that the rates are more easily explained by AGN activity in the past than driving by star formation. So we should not expect every galaxy with an outflow to obey (111) perfectly. As another test of the picture given here, AGN close to their Eddington luminosities (so that $L \propto M \propto \sigma^4$ and $l \simeq 1$), should show clearout rates $\propto \sigma^{8/3}$ (Eq. 107) scaling linearly with the driving luminosity L. Figure 9 of Cicone et al. [8] tends to support this correlation, with about the expected normalization.

8.5 SMBH Feedback in General

The bulge feedback we discussed above has several other consequences. As we have seen, as the SMBH reaches the M_σ mass it drives a pressure wave through the bulge gas. Much of this sweeps up and drives the interstellar gas, pushing this to the edge of the galaxy where it meets and resists the inflow of primordial cosmological gas. The important feature here is that the bulge gas has itself been through generations of stars, so this effect is what enriches the circumgalactic medium with metals.

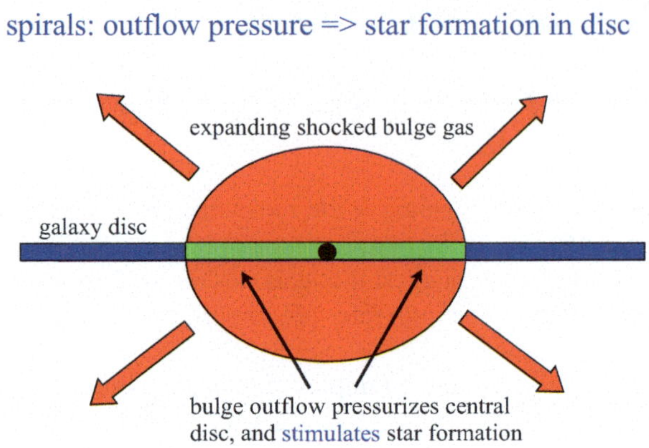

Fig. 33 Star formation in spiral galaxies driven by the expulsion of the bulge gas

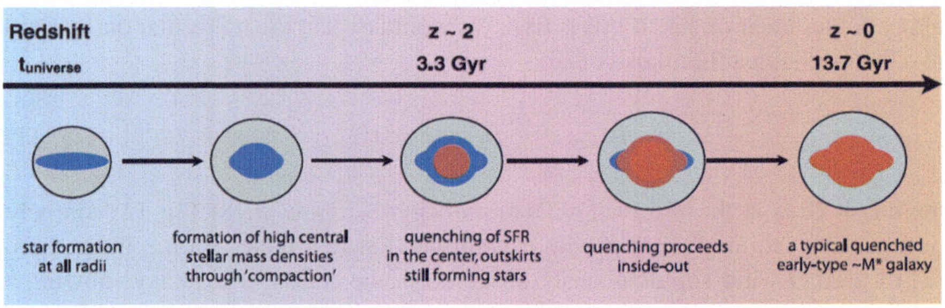

Fig. 34 Schematic picture of the evolution of star formation in spiral galaxies

The other effect of the pressure wave is of course that gas with densities close to that needed for star formation is induced to form stars as it passes—this is often called positive feedback, as opposed to the negative effects of sweep-out we have discussed. This effect goes further—as the gas is pushed out of the bulge, it travels outwards over both faces of the galaxy disc and can in principle cause a wave of inside-out star formation here (Figs. 33 and 34).

This wave does not cross the disc with the constant speed v_e (Eq. 98), but more slowly, because the AGN driving must be variable [76] (Fig. 35).

Schlegel et al. (2016) show an example where the SMBH appears to have launched two successive outflows, visible as X-ray arcs outside the main body of the galaxy. We might also ask whether any such large-scale effects of SMBH feedback are seen in our own galaxy. The answer appears to be positive—the so-called Fermi bubbles, seen in gamma-rays by the Fermi-LAT experiment (Fig. 36).

This detected two symmetrical gamma-ray bubbles each side of the plane of the Milky Way. One suggestion is that the bubbles are roughly steady structures blown by star formation. Zubovas et al. [79] suggested instead that they could be the remains of a wide-angle outflow from Sgr A*. This would be a natural result of the star

Fig. 35 X-ray arcs around a galaxy may reveal repeating blow-out episodes

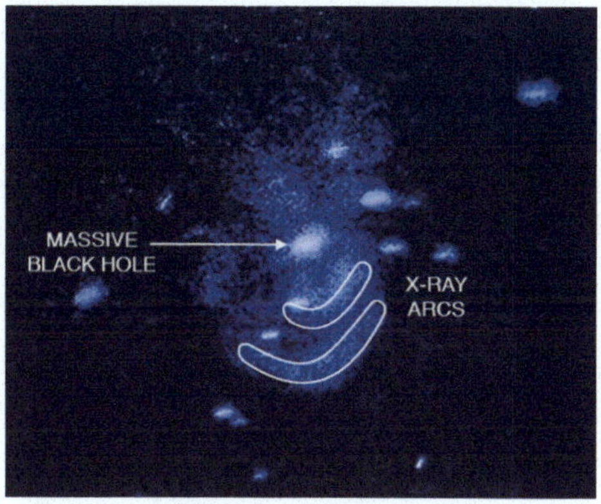

(Schlegel et al., 2016)

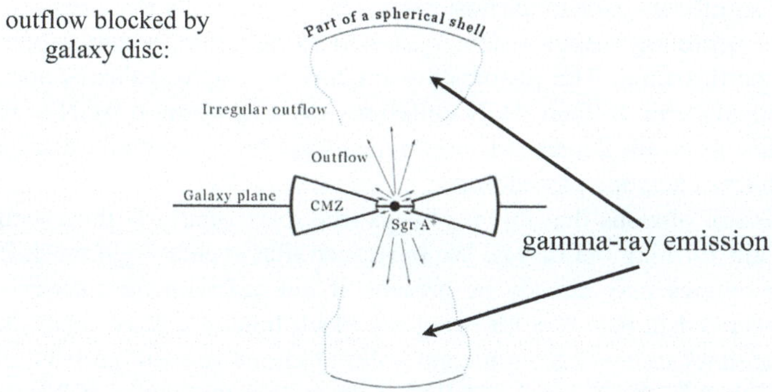

Fig. 36 The Fermi bubbles of the Milky Way

formation event in the inner 0.5 pc of the Milky Way 6 Myr ago which left the well-observed families of orbiting stars we discussed in Sect. 1 above. The outflow is quasispherical close to the SMBH, but forced into a pair of symmetrical lobes by the gas pressure of the Galactic plane, and shocks against the interstellar gas in the Galaxy bulge. The gamma-rays emission could come from cosmic rays created either directly by the SMBH or in the shocks. The outflow would not affect the plane of the Galaxy, agreeing with the considerable observational evidence that SMBH properties do not correlate with galaxy disc properties. The estimated accreted mass powering the event and so the Fermi lobes is $\sim 2 \times 10^3 M_\odot$, close to the mass of young stars formed in the event, and suggesting that similar amounts of infalling gas formed stars or got accreted. Of course this event is tiny on the scale of those powering significant SMBH growth and establishing the $M-\sigma$ relation, but it corresponds to the hot phase of the energy-driven outflows discussed in this section. This suggests that future gamma-ray observatories might detect similar lobes associated with the powerful molecular outflows currently found in distant galaxies. If this explanation is correct we expect the lobes to be symmetrical about the galaxy disc plane, as in the Milky Way.

Another suggested mechanism for the Fermi bubbles is driving by jets from the central black hole. But we have already seen that observed jet directions are in general completely uncorrelated with the galaxy structure, so the symmetry of the Fermi bubbles is a problem for this approach.

8.6 Radiation Feedback

So far all the feedback effects I have discussed have been driven by the mechanical luminosity $\sim(\eta/2)L_{\rm Edd} \sim 0.05L_{\rm Edd}$ of the black hole wind. As we saw, even this had to be reduced by a further efficiency factor $\sim\sigma/c \sim 10^{-3}$ to avoid the danger of dispersing significant parts of the host galaxy. We have largely ignored the remaining 95% of the accretion output which is emitted as radiation, which is potentially a far larger perturbation. The justification for this is simple: galaxies appear to be largely optically thin to their AGN emission—we can observe AGN at almost all wavelengths, after all. Direct radiation appears far less potentially disruptive than even much less energetic gas outflows.

But it is not obvious that this is always true, particularly at the epoch when a galaxy is still forming out of gas. We have seen that even at high redshift, SMBH with large masses may already be present, if not yet with the masses $\sim10^9{\rm M}_\odot$ seen at lower redshifts $z \sim 6$. Growing the black hole in a reasonable time needs a large gas mass to cover a significant solid angle close to it, so it is reasonable to imagine that an unobscured SMBH has moved around some of this gas in the past. The radiation field is by far the largest perturbation an AGN makes on its environment. So it is likely that radiation forces push the gas outwards, spreading it over a larger solid angle and reducing its optical depth along the line of sight. As they make their way outwards, the AGN photons scatter slightly inelastically so that the AGN luminosity L does work lifting the gas. Assuming this was not in large-scale dynamical motion before being pushed outwards, we take it to have an isothermal distribution, obeying (88). Radiation pressure progressively sweeps the innermost gas outwards with undisturbed gas outside it. King [23] show that this implies a total electron scattering optical depth

$$\tau_{\rm tot}(R) = \tau(R) + \tau_{\rm sh}(R) \simeq \frac{\kappa f_g \sigma^2}{\pi G R}. \tag{112}$$

For small R the gas is very optically thick, so the accretion luminosity of the SMBH is initially largely trapped and isotropized by scattering, and there is a growing radiation pressure inside R. This lifts the gas against its weight

$$W(R) = \frac{GM(R)M_g(R)}{R^2} = \frac{4 f_g \sigma^4}{G} \tag{113}$$

King [23] show that the inner radius R obeys an equation very like the equation of motion (96) for energy-driving by a shocked but adiabatic black hole wind, but with effective adiabatic index $\gamma = 4/3$ appropriate for radiation rather than $\gamma = 5/3$ for a monatomic gas. The attractor solution corresponding to (98) obeys

$$L = \frac{4 f_g \sigma^2 v_e^3}{G} + \frac{12 f_g \sigma^4}{G} v_e. \tag{114}$$

Since

$$L = \frac{\mathrm{d}E}{\mathrm{d}R} v_e \tag{115}$$

where E is the total radiation energy inside R, we can write

$$\frac{\mathrm{d}E}{\mathrm{d}R} = \left(3 + \frac{v_e^2}{\sigma^2}\right) W \tag{116}$$

It is clear that radiation driving effectively stops once the gas becomes optically thin to radiation. This defines a 'transparency radius'

$$R_{\mathrm{tr}} \sim \frac{\kappa f_g \sigma^2}{\pi G} \simeq 50 \left(\frac{f_g}{0.16}\right) \sigma_{200}^2 \ \mathrm{pc}, \tag{117}$$

where (up to a logarithmic factor) the optical depth $\tau_{\mathrm{tot}} \sim 1$. Radiation now begins to leak out, and the outward radiation driving effectively stops, so $v_e \sim 0$. Equation (116) now shows that the total accretion energy needed to drive the gas originally surrounding the black hole to the transparency radius R_{tr} is

$$E_{\mathrm{tr}} \simeq 3 W R_{\mathrm{tr}} = \frac{12 \kappa f_g^2 \sigma^6}{\pi G^2}, \tag{118}$$

which requires only a tiny accreted mass

$$\Delta M \gtrsim \frac{E_{\mathrm{tr}}}{\eta c^2} \sim 3 \times 10^3 \sigma_{200}^6 \mathrm{M}_{\odot}, \tag{119}$$

far smaller than the final SMBH mass. This means that the surroundings of SMBH tend to become transparent very early in their lives. The gas driven out to this radius has mass

$$M(R_{\mathrm{tr}}) = 2 f_g \frac{f_g \kappa \sigma^4}{\pi G^2} = 2 f_g M_\sigma \sim M_\sigma \tag{120}$$

comparable to the final SMBH mass M_σ. This stalling outflow now constitutes a significant obstacle to later outflows from the SMBH. For SMBH mass below M_σ these are momentum-driven UFOs, which therefore shock at distances $\sim R_{\mathrm{tr}}$. This gas may be what is identified as 'warm absorbers' in X-ray spectra of AGN.

9 The Black Hole—Bulge Mass Relation

The relation between black hole mass M and the total stellar mass of the host galaxy bulge M_b was historically the first relation found between the properties of supermassive black holes and their hosts [42, 46]. As we have seen, the existence of any relation of this type was initially a surprise, given how insignificant the direct gravitational effect of an SMBH is.

One approach [20, 55] simply asserts that SMBH and their hosts grow simultaneously through repeated mergers of smaller galaxies with uncorrelated M, M_b. Then the central limit theorem gradually produces a proportionality between M and M_b, with scatter decreasing for higher M, M_b (although the value of M/M_b is left undetermined by this process).

But we know that the SMBH binding energy is significant, making it very likely that the black hole mass is set because its own feedback limits the mass reservoir is can call on for growth. We have seen above how very simple arguments using this idea give the $M-\sigma$ relation remarkably accurately, without a real free parameter, and almost every attempt at deriving this relation has used this argument.

Then it is not plausible that the $M-M_b$ relation results from SMBH feedback. If we accept that M is fixed by the gas mass distribution encoded in σ, we cannot simultaneously argue that M_b somehow directly fixes M. And the reverse argument that the black hole mass alone determines the stellar bulge mass does not work, as we have seen that the SMBH has no direct gravitational effect on the the large-scale bulge, and influences it only by driving away the remaining gas after reaching M_σ.

This last point suggests the real nature of the M, M_b connection. The bulge stellar mass is not affected by the SMBH, so must be proportional to σ^4 quite independently of the black hole. This is known observationally: the Faber–Jackson relation [15] for elliptical galaxies (which are in a sense 'all bulge') asserts that the total stellar luminosity L_* obeys

$$L_* \sim 2 \times 10^{10} \mathrm{L}_\odot \text{œ}_{200}^4, \tag{121}$$

which for stellar mass-to-light ratios $\sim$5 gives

$$M_* \sim 1 \times 10^{11} \mathrm{M}_\odot \text{œ}_{200}^4 \tag{122}$$

Comparing with (93) we see directly that

$$M \sim 10^{-3} M_b \tag{123}$$

So the main question remaining is the physical origin of the Faber–Jackson relation (121). The obvious possibility is that just as black hole wind feedback fixes $M-\sigma$, stellar feedback, through stellar winds and supernovae, fixes $M_b - \sigma$ in a parallel way. This idea (cf Murray et al. [49]) is widely accepted. Power et al. [60] use this approach to get the relation

$$M_b \sim \frac{0.14 f_g t_H \sigma^4}{h(z)\epsilon_* c G}. \tag{124}$$

Here $\epsilon_* \simeq 2 \times 10^{-3}$ is the total luminous energy yield from a main-sequence star in terms of its rest-mass energy $M_* c^2$, t_H is the Hubble time, and $h(z)$ is the dimensionless Hubble factor at redshift z. The latter dependence arises because at high redshift the galaxy bulge grows over time by accreting primordial gas. Comparing with (91) gives

$$M \simeq M_\sigma \sim \frac{1.8\kappa\epsilon_* c h(z)}{\pi G t_H} M_b \sim 10^{-3} M_b, \tag{125}$$

where the last relation is for zero redshift ($h(z) = 1$) (M/M_b is larger at high redshift). The proportionality factor $\sim 10^{-3}$ between M and M_b arises because effective black-hole feedback per unit mass is about 1000 times larger than effective stellar feedback per unit mass. In this view the $M - M_b$ connection is acausal—that is, there is no physics directly linking M and M_b, but instead two parallel relations with σ. It is these that are fundamental, rather than the $M - M_b$ connection.

10 Conclusion

I hope these notes have made clear that the subject of supermassive black hole growth, and its connection to the evolution of the host galaxies, has made rapid progress over recent years. The discovery of the $M-\sigma$ relation at the beginning of this century greatly stimulated this. But big questions remain. Probably the most important is to discover what properties of the host determine how mass accretes on to the central supermassive black hole. Without a clear quantitative picture of this, the subject remains partially indeterminate. As a historical analogy, it probably has a similar status to stellar evolution theory before astronomers understood nuclear burning: then as now, they made plausible guesses, and were still able to make progress. But the arrival of a full picture of nucleosynthesis transformed the subject completely, and gave a satisfying picture of how stars evolve which is still developing. We should be encouraged by this: there are exciting times ahead.

Acknowledgements I thank the organisers, particularly Roland Walter, Nicolas Produit and Lucio Mayer, for making this school so enjoyable, and for valuable comments on the manuscript. I thank the students and my fellow lecturers and all the other participants for their enthusiasm, interest, and their many questions.

References

1. Abramowicz, M.A., Czerny, B., Lasota, J.P., Szuszkiewicz, E.: Slim accretion disks. ApJ **332**, 646 (1988)

2. Assef, R.J., et al.: Half of the most luminous quasars may be obscured: investigating the nature of WISE-selected hot dust-obscured galaxies. ApJ **804**, 27. arXiv:1408.1092 [astro-ph.GA] (2015)
3. Balbus, S.A., Hawley, J.F.: A powerful local shear instability in weakly magnetized disks. I. Linear analysis. ApJ **376**, 214 (1991)
4. Bardeen, J.M., Petterson, J.A.: The lense-thirring effect and Accretion disks around Kerr black holes. ApJ **195**, L65 (1975)
5. Barth, A.J., Martini, P., Nelson, C.H., Ho, L.C.: Iron Emission in the z = 6.4 Quasar SDSS J114816.64+525150.3. ApJ **594**, L95–L98. arXiv:astro-ph/0308005 [astro-ph] (2003)
6. Bondi, H.: On spherically symmetrical accretion. MNRAS **112**, 195 (1952)
7. Chandrasekhar, S.: Dynamical friction. I. General considerations: the co-efficient of dynamical friction. ApJ **97**, 255 (1943)
8. Cicone, C., et al.: Massive molecular outflows and evidence for AGN feedback from CO observations. A& A **562**, A21. arXiv:1311.2595 [astro-ph.CO] (2014)
9. Collin-Souffrin, S., Dumont, A.M.: Line and continuum emission from the outer regions of accretion discs in active galactic nuclei. II. Radial structure of the disc. A& A **229**, 292–301 (1990)
10. Costa, T., Sijacki, D., Haehnelt, M.G.: Feedback from active galactic nuclei: energy- versus momentum-driving. MNRAS **444**, 2355–2376. arXiv:1406.2691 [astro-ph.GA] (2014)
11. Dehnen, W., King, A.: Black hole foraging: feedback drives feeding. ApJ **777**, L28. arXiv:1310.2039 [astrospsph.CO] (2013)
12. Denney, K.D., et al.: A revised broad-line region radius and black hole mass for the narrow-line Seyfert 1 NGC 4051. ApJ **702**, 1353–1366. arXiv:0904.0251 [astro-ph.CO] (2009)
13. Dyson, J.E., Williams, D.A.: The Physics of the Interstellar Medium. Bristol Institute of Physics Publishing, Bristol (1997)
14. European Southern Observatory (2019)
15. Faber, S.M., Jackson, R.E.: Velocity dispersions and mass-to-light ratios for elliptical galaxies. ApJ **204**, 668–683 (1976)
16. Faucher-Giguère, C.-A., Quataert, E.: The physics of galactic winds driven by active galactic nuclei. MNRAS **425**, 605–622. arXiv:1204.2547 [astro-ph.CO] (2012)
17. Frank, J., King, A.R., Raine, D.J.: Accretion power in astrophysics (2002)
18. Ghisellini, G., et al.: Chasing the heaviest black holes of jetted active galactic nuclei. MNRAS **405**, 387–400. arXiv:0912.0001 [astro-ph.HE] (2010)
19. Hawking, S.W.: Black holes in general relativity. Commun. Math. Phys. **25**, 152–166 (1972)
20. Jahnke, K., Macciò, A.V.: The non-causal origin of the Black-holegalaxy scaling relations. ApJ **734**, 92. arXiv:1006.0482 [astro-ph.CO] (2011)
21. King, A.: Black holes, galaxy formation, and the $M_{BH} - \sigma$ relation. ApJ **596**, L27–L29. arXiv:astro-ph/0308342 [astro-ph] (2003)
22. King, A.: The AGN-starburst connection, galactic superwinds, and $M_{BH} - \sigma$. ApJ **635**, L121–L123. arXiv:astro-ph/0511034 [astro-ph] (2005)
23. King, A.: The Supermassive Black Hole-Galaxy connection. Space Sci. Rev. **183**, 427–451 (2014)
24. King, A.: How big can a black hole grow? MNRAS **456**, L109–L112. arXiv:1511.08502 [astro-ph.GA] (2016)
25. King, A., Muldrew, S.I.: Black hole winds II: Hyper-Eddington winds and feedback. MNRAS **455**, 1211–1217. arXiv:1510.01736 [astro-ph.HE] (2016)
26. King, A., Nixon, C.: AGN flickering and chaotic accretion. MNRAS **453**, L46–L47. arXiv:1507.05960 [astro-ph.HE] (2015)
27. King, A., Nixon, C.: Black holes in stellar-mass binary systems: expiating original spin? MNRAS **462**, 464–467. arXiv:1607.02144 [astro-ph.HE] (2016)
28. King, A., Nixon, C.: Misaligned accretion and jet production. ApJ **857**, L7. arXiv:1803.08927 [astro-ph.HE] (2018)
29. King, A., Pounds, K.: Powerful outflows and feedback from active galactic nuclei. ARA& A **53**, 115–154. arXiv:1503.05206 [astro-ph.GA] (2015)

30. King, A.R.: Masses, beaming and Eddington ratios in ultraluminous X-ray sources. MNRAS **393**, L41–L44. arXiv:0811.1473 [astro-ph] (2009)
31. King, A.R., Lasota, J.P.: Magnetic alignment of rotating black holes and accretion discs. A& A **58**, 175–179 (1977)
32. King, A.R., Lubow, S.H., Ogilvie, G.I. Pringle, J.E.: Aligning spinning black holes and accretion discs. MNRAS **363**, 49–56. arXiv:astro-ph/0507098 [astro-ph] (2005)
33. King, A.R., Pringle, J.E.: Growing supermassive black holes by chaotic accretion. MNRAS **373**, L90–L92. arXiv:astro-ph/0609598 [astro-ph] (2006)
34. King, A.R., Pringle, J.E.: Fuelling active galactic nuclei. MNRAS **377**, L25–L28. arXiv:astro-ph/0701679 [astro-ph] (2007)
35. King, A.R., Pringle, J.E., Livio, M.: Accretion disc viscosity: how big is alpha? MNRAS **376**, 1740–1746. arXiv:astro-ph/0701803 [astro-ph] (2007)
36. King, A.R., Pringle, J.E., Hofmann, J.A.: The evolution of black hole mass and spin in active galactic nuclei. MNRAS **385**, 1621–1627. arXiv:0801.1564 [astro-ph] (2008)
37. King, A.R., Pounds, K.A.: Black hole winds. MNRAS **345**, 657–659. arXiv:astro-ph/0305541 [astro-ph] (2003)
38. King, A.R., Ritter, H.: Cygnus X-2, super-Eddington mass transfer, and pulsar binaries. MNRAS **309**, 253–260. arXiv:astro-ph/9812343 [astro-ph] (1999)
39. King, A.R., Zubovas, K., Power, C.: Large-scale outflows in galaxies. MNRAS **415**, L6–L10. arXiv:1104.3682 [astro-ph.GA] (2011)
40. Kinney, A.L., et al.: Jet Directions in seyfert galaxies. ApJ **537**, 152–177. arXiv:astro-ph/0002131 [astro-ph] (2000)
41. Kormendy, J., Ho, L.C.: Coevolution (Or Not) of supermassive black holes and host galaxies. ARA& A **51**, 511–653. arXiv:1304.7762 [astro-ph.CO] (2013)
42. Kormendy, J., Richstone, D.: Inward bound—the search for supermassive Black Holes in galactic nuclei. ARA& A **33**, 581 (1995)
43. Lodato, G., Price, D.J.: On the diffusive propagation of warps in thin accretion discs. MNRAS **405**, 1212–1226. arXiv:1002.2973 [astro-ph.HE] (2010)
44. Lynden-Bell, D.: Galactic nuclei as collapsed old quasars. Nature **223**, 690–694 (1969)
45. Lynden-Bell, D., Pringle, J.E.: The evolution of viscous discs and the origin of the nebular variables. MNRAS **168**, 603–637 (1974)
46. Magorrian, J., et al.: The demography of massive dark objects in galaxy centers. AJ **115**, 2285–2305. arXiv:astro-ph/9708072 [astro-ph] (1998)
47. Mayer, L.: Super-Eddington accretion; flow regimes and conditions in high-z galaxies. arXiv:1807.06243 [astro-ph.HE] (2018)
48. McQuillin, R.C., McLaughlin, D.E.: Momentum-driven feedback and the M-σ relation in non-isothermal galaxies. MNRAS **423**, 2162–2176. arXiv:1204.2082 [astro-ph.CO] (2012)
49. Murray, N., Quataert, E., Thompson, T.A.: ApJ **618**, 569 (2005)
50. Nayakshin, S. Two-phase model for black hole feeding and feedback. MNRAS **437**, 2404–2411. arXiv:1311.4492 [astro-ph.GA] (2014)
51. Nixon, C., King, A., Price, D., Frank, J.: Tearing up the disk: how black holes accrete. ApJ **757**, L24. arXiv:1209.1393 [astro-ph.HE] (2012)
52. Nixon, C.J., King, A.R.: Broken discs: warp propagation in accretion discs. MNRAS **421**, 1201–1208. arXiv:1201.1297 [astro-ph.HE] (2012)
53. Ohsuga, K., Mineshige, S., Mori, M., Umemura, M.: Does the slim-disk model correctly consider photon-trapping effects? ApJ **574**, 315–324. arXiv:astro-ph/0203425 [astro-ph] (2002)
54. Papaloizou, J.C.B., Pringle, J.E.: The time-dependence of non-planar accretion discs. MNRAS **202**, 1181–1194 (1983)
55. Peng, C.Y.: How mergers may affect the mass scaling relation between gravitationally bound systems. ApJ **671**, 1098–1107. arXiv:0704.1860 [astro-ph] (2007)
56. Pontzen, A., Governato, F.: How supernova feedback turns dark matter cusps into cores. MNRAS **421**, 3464–3471. arXiv:1106.0499 [astro-ph.CO] (2012)
57. Pounds, K.A., King, A.R.: The shocked outflow in NGC 4051 - momentumdriven feedback, ultrafast outflows and warm absorbers. MNRAS **433**, 1369–1377. arXiv:1305.2046 [astro-ph.HE] (2013)

58. Pounds, K.A., Nixon, C.J., Lobban, A., King, A.R.: An ultrafast inow in the luminous Seyfert PG1211+143. MNRAS **481**, 1832–1838. arXiv:1808.09373 [astro-ph.HE] (2018)
59. Pounds, K.A., Vaughan, S.: An extended XMM-Newton observation of the Seyfert galaxy NGC 4051 - I. Evidence for a shocked outflow. MNRAS **413**, 1251–1263. arXiv:1012.0998 [astro-ph.CO] (2011)
60. Power, C., Zubovas, K., Nayakshin, S., King, A.R.: MNRAS **413**, 110 (2011)
61. Richings, A.J., Faucher-Giguère, C.-A.: The origin of fast molecular outflows in quasars: molecule formation in AGN-driven galactic winds. MNRAS **474**, 3673–3699. arXiv:1706.03784 [astro-ph.GA] (2018a)
62. Richings, A. J., Faucher-Giguère, C.-A.: Radiative cooling of swept-up gas in AGN-driven galactic winds and its implications for molecular outflows. MNRAS **478**, 3100–3119. arXiv:1710.09433 [astro-ph.GA] (2018b)
63. Salpeter, E.E.: Accretion of interstellar matter by massive objects. ApJ **140**, 796–800 (1964)
64. Schawinski, K., Koss, M., Berney, S., Sartori, L. F.: Active galactic nuclei flicker: an observational estimate of the duration of black hole growth phases of 105 yr. MNRAS **451**, 2517–2523. arXiv:1505.06733 [astrospsph.GA] (2015)
65. Scheuer, P.A.G., Feiler, R.: The realignment of a black hole misaligned with its accretion disc. MNRAS **282**, 291 (1996)
66. Shakura, N.I., Sunyaev, R.A.: Reprint of 1973A & A....24..337S. Black holes in binary systems. Observational appearance. A& A **500**, 33– 51 (1973)
67. Silk, J., Nusser, A.: The Massive-black-hole-velocity-dispersion relation and the Halo Baryon fraction: a case for positive active Galactic nucleus feedback. ApJ **725**, 556–560. arXiv:1004.0857 [astro-ph.CO] (2010)
68. Soltan, A.: Masses of quasars. MNRAS **200**, 115–122 (1982)
69. Tombesi, F., et al.: Evidence for ultra-fast outflows in radio-quiet AGNs. I. Detection and statistical incidence of Fe K-shell absorption lines. A& A **521**, A57. arXiv:1006.2858 [astro-ph.HE] (2010)
70. Tombesi, F., et al.: Evidence for ultra-fast outflows in radio-quiet active galactic nuclei. II. Detailed photoionization modeling of Fe K-shell absorption lines. ApJ **742**, 44. arXiv:1109.2882 [astro-ph.HE] (2011)
71. Tombesi, F., et al.: Wind from the black-hole accretion disk driving a molecular outflow in an active galaxy. Nature **519**, 436–438. arXiv:1501.07664 [astro-ph.HE] (2015)
72. Toomre, A.: On the gravitational stability of a disk of stars. ApJ **139**, 1217–1238 (1964)
73. Walker, S.A., Fabian, A.C., Russell, H.R., Sanders, J.S.: The effect of the quasar H1821+643 on the surrounding intracluster medium: revealing the underlying cooling flow. MNRAS **442**, 2809–2816. arXiv:1405.7522 [astro-ph.HE] (2014)
74. Willott, C.J., McLure, R.J., Jarvis, M.J.: A $3\times10^9 M_{solar}$ black hole in the quasar SDSS J1148+5251 at z=6.41. ApJ **587**, L15–L18. arXiv:astro-ph/0303062 [astro-ph] (2003)
75. Zubovas, K., King, A.: BAL QSOs and extreme UFOs: the Eddington connection. ApJ **769**, 51. arXiv:1304.1691 [astro-ph.GA] (2013)
76. Zubovas, K., King, A.: The small observed scale of AGN-driven outflows, and inside-out disc quenching. MNRAS **462**, 4055–4066. arXiv:1607.07258 [astro-ph.GA] (2016)
77. Zubovas, K., King, A.R.: The M-σ relation in different environments. MNRAS **426**, 2751–2757. arXiv:1208.1380 [astro-ph.GA] (2012)
78. Zubovas, K., King, A.R.: Galaxy-wide outflows: cold gas and star formation at high speeds. MNRAS **439**, 400–406. arXiv:1401.0392 [astro-ph.GA] (2014)
79. Zubovas, K., King, A.R., Nayakshin, S.: The Milky Way's Fermi bubbles: echoes of the last quasar outburst? MNRAS **415**, L21–L25. arXiv:1104.5443 [astro-ph.GA] (2011)

Black Holes Across Cosmic History: A Journey Through 13.8 Billion Years

Tiziana Di Matteo

Contents

T. Di Matteo (✉)
McWilliams Center for Cosmology, Carnegile Mellon University, Pittsburgh, PA 15213, USA
e-mail: tiziana@phys.cmu.edu

School of Physics, The University of Melbourne, VIC 3010, Australia

R. Walter et al. (eds.), *Black Hole Formation and Growth*, Saas-Fee Advanced Course 48,
https://doi.org/10.1007/978-3-662-59799-6_3

Abstract Massive black holes are fundamental constituents of our cosmos, from the Big Bang to today. Understanding their formation at cosmic dawn, their growth, and the emergence of the first, rare quasars in the early Universe remains one of our greatest theoretical and observational challenges. Hydrodynamic cosmological simulations self-consistently combine the processes of structure formation at cosmological scales with the physics of smaller, galaxy scales. They capture our most realistic understanding of massive black holes and their connection to galaxy formation and have become the primary avenue for theoretical research in this field. The space-based gravitational wave telescope LISA will open up new investigations into the dynamical processes involving massive black holes. Multi-messenger astrophysics brings new exciting prospects for tracing the origin, growth and merger history of massive black holes across cosmic ages.

In this chapter we will take a journey through our cosmic history to examine the role of black holes from the Big Bang to today. Black holes are fundamental components of our Universe, and they play a major role in our understanding of galaxy formation. We will discuss how primordial black holes may form during the Big Bang, from the initial density inhomogeneities set at the epoch of Inflation. We will see how either these objects or black holes forming from the collapse of the first density peaks and associated processes at cosmic dawn (the first epoch of galaxy formation) are likely to lead to a significant population of 'seed' black holes that merge and grow. We will examine the emergence of the first population of supermassive black holes, the first quasars, that occurs within the first billion years of our cosmic history.

As we link the formation of the first black holes to the first quasars, we also study the formation of black holes in our standard paradigm of structure formation. We will describe how we understand structure formation within the context of cosmological simulations. State-of-the art cosmological simulations include the formation and growth of black holes and can make direct predictions for current and future observations. We will discuss the growth of black holes in the centers of galaxies, and how mergers and gas accretion induced by large inflows and mergers connect the physics of small scales to larger cosmological scales. Cosmological simulations allow us to study directly the co-evolution of and connection between black hole growth and galaxy formation and the emergence of the fundamental relations between central black holes and their host galaxy properties. Figure 1 schematically summarizes the sequence of topics that we will cover, dealing with massive black holes from the early universe to today.

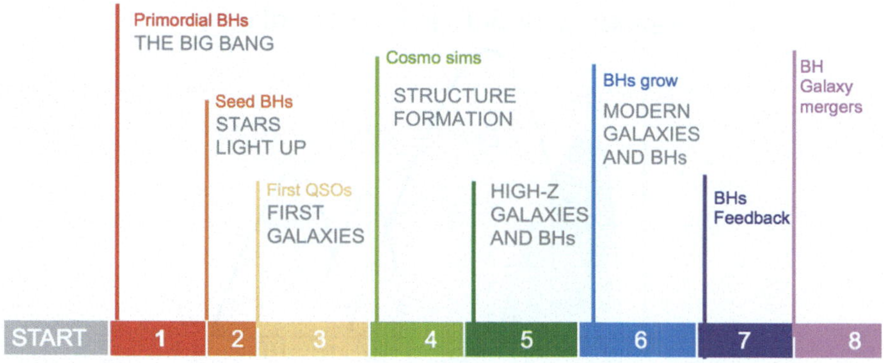

Fig. 1 Schematic outline of the flow of this paper, Black Holes Across Cosmic History. The paper is subdivided into chapters which follow the evolution of black holes from the Big Bang to today. The numbers corresponds to the sections in the text

1 Primordial Black Holes: Forming Black Holes During Inflation

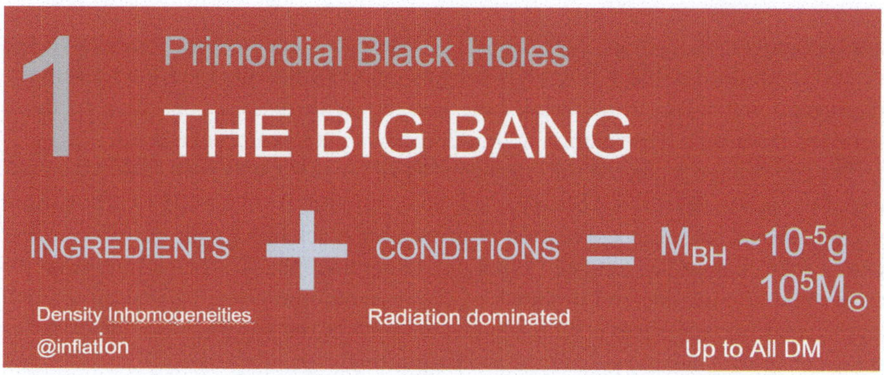

So let's get started with primordial black holes: this is a speculative topic, as there is no evidence for the existence of this black hole population. It has been theorized and discussed for the last 50 years and this topic keeps resurfacing. It is therefore worth reviewing the possible origin of a primordial BH population.

Primordial Black Holes are a population of BHs that is speculated to emerge during the collapse of the first density fluctuations. The fundamental ingredient for the emergence of primordial black holes is the formation of density fluctuations during the epoch of inflation and their collapse during the radiation dominated era of our Universe [1–5]. Predictions imply that a population of black holes could exist ranging in mass from 10^{-5} grams up to 10000 solar masses. The most significant consequence of a population of primordial black holes in our Universe is that it could make up a

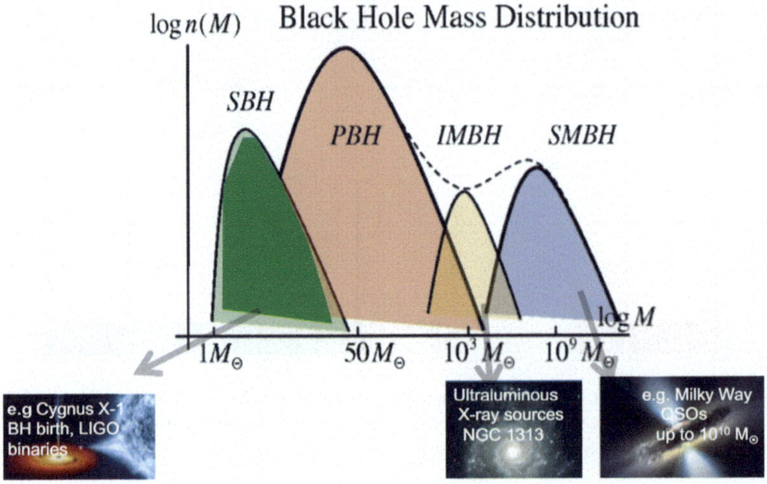

Fig. 2 Schematic illustration of the black hole mass distribution. There is solid evidence for the population of Super Massive Black Holes (SMBH), Stellar Black Holes (SBH) and some circumstantial evidence for Intermediate Mass Black holes. The mass range predicted for PBH is not yet ruled out

large component of all of the dark matter [1]. While this remains fully speculative, the interest in primordial black holes has been re-ignited following the LIGO discovery of gravitational waves from black hole mergers with masses consistent to those theorised from such a population and not previously known/observed [6].

The mass range predicted for primordial blackholes fits very well within the context of the newly discovered binary black holes inferred from the LIGO observation of gravitational waves. A population of primordial black holes with masses of a few tens of solar masses has not been ruled out. Other astrophysical black holes that we know about include the end products of stellar evolution (a few solar masses), and much larger masses, up to a billion solar masses (for the supermassive black holes in the center of galaxies) (see Fig. 2 as illustration).

The black holes detected by LIGO correspond to a new population of black holes which have not been observed before. While intermediate-mass black supermassive black holes are known to populate the centres of globular clusters and galaxies respectively this new class of black holes in binaries had not been detected previously. Primordial black holes in this mass range are poorly constrained, and could still exist in substantial number in the Universe today and even potentially constitute the bulk of dark matter [1]. When two such primordial black holes get sufficiently close in the galactic halo they would indeed radiate enough gravitational radiation to become bound and eventually inspiral and give rise to the gravitational wave signal that we observe.

Primordial black holes have long been a source of intense interest despite no evidence for their existence. They are relevant in many different contexts, and are very interesting objects, small enough for Hawking radiation to be important. This is why they were discussed in the early days when Hawking radiation was first predicted. Hawking radiation [7] still remains one of the key features of black holes which unite thermodynamics with general activity and quantum mechanics. It would be a revolutionary discovery to find evidence for their existence, as they probe an epoch of the early universe which cannot otherwise be probed. This is something that makes primordial black holes extremely appealing. Primordial black holes with masses less than 10^{15} grams are those that have already evaporated by now but the ones that have a larger masses (up to those currently inferred by LIGO) would still be in existence as they would have not evaporated. This population is potentially very exciting as it could constitute about 25% of the critical density of the Universe, perhaps enough to constitute a large proportion of the dark matter [1].

These ideas go back to the earliest days of PBH research which looked at their formation from the collapse of primordial density inhomogeneities left from Inflation during the radiation area [2, 5] (see Fig. 3 for a schematic timeline (taken from a review by Garcia-Bellido and Clesse in Scientific American [4]). Although other processes at this early epoch have also been hypothesized for the formation of PBHs e.g., [8–10]. Because of this, PBH are not subject to the Big Bang nucleosynthesis constraint (which is that baryonic matter can make up only ~5% of the critical density). The remnant of this primordial population, if not evaporated, could be massive enough to be BH seeds themselves for some of the supermassive black holes and galaxies formed after the Big Bang. Today primordial black hole clusters could orbit in the halos of massive galaxies [11]. There is no direct evidence that PBHs are the DM but there is also none for traditional CDM candidates. WIMPS (weakly interacting massive particles) have been rather elusive: 30 yrs of accelerator experiments and direct dark matter searches have found nothing. In the 1990 there was flurry of activity searching for alternative dark matter candidates, MACHOs (massive astrophysical compact objects). Microlensing studies though showed that they could constitute at most 20% of DM. As an example of specific MACHO candidates, there are objects predicted to have masses of $\sim$0.5 $M_\odot$ which result from processes occurring during the QCD phase transition. However, black holes that potentially result from collapse of density inhomogeneities at the radiation epoch could be produced in copious amounts, and for certain parameters these black holes may constitute the dark matter in the Universe.

If one writes down the expression of the Schwarzchild radius of black hole, in terms of a density this yields $\rho_s = 10^{18}(M\,M_\odot)^{-2}$ g/cm^{-3}. It is easy to see than that in today's Universe black holes can never be smaller than about 1 $M_\odot$. However if we increase the cosmological density, such as expected at early times we can find the connection between the PBH mass and horizon mass at formation. During the radiation dominated era $\rho \propto a(t){-}4$ where $a(t)$ is the scale factor, which scales as $t^{-1/2}$ at radiation dominated era. Using this expression we find that $\rho \sim (1/Gt^2) \sim 10^6(t/1s)^2$ g/cm^{-3}. Finally, during the radiation era the sound speed is of the order of $c/\sqrt{3}$ and hence $R_s \sim R_{Jeans} \sim R_h$. Any linear perturbation collapses at a critical overdensity of 0.1–0.7 so that a fraction less than 1 goes into the formation of a PBH. High enough densities are a necessary but not sufficient condition for PBH formation. This will depend on the specific form of the potential at the time of Inflation, and

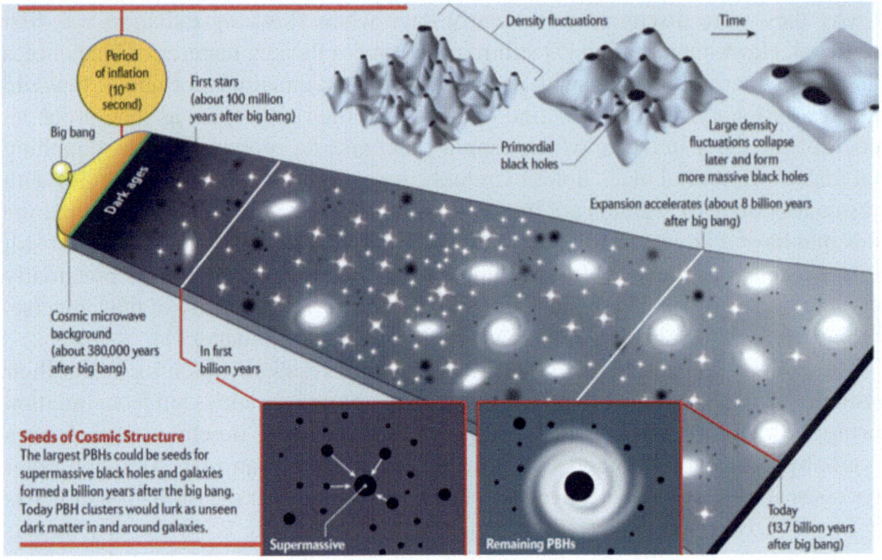

Fig. 3 PBH are created during the period of inflation (yellow region in the diagram) of cosmic history. Depending on the details of the inflation scenario, if their collapse occurs from the initial density perturbations in the radiation dominated epoch they can be massive enough to survive in the halos of present day galaxies. Adapted from [4]

hence on the particular Inflation model. In the most commonly discussed inflation models, large overdense regions could stop expanding and collapse. For $t = 1$ s, if this collapse occurs, primordial BHs, whose mass then goes as $M_{PBH} = c^3 t/G$ can attain masses of $10^5\,M_{\odot}$. PBH of this mass can form from critical collapse of Gaussian perturbations in the radiation dominated era. Even with a tiny collapsed fraction of PBHs during radiation era it would be easy to produce all the DM [1].

So in summary PBH are extremely interesting objects as they probe physics on scales that are otherwise inaccessible by observations. The dark matter can be made of PBHs. They produce MACHOs, IMBHs, ULX, and could perhaps be the seeds for the SMBHs in galaxies today.

The current parameter space for the PBH is summarised in Fig. 3. It has been constrained by MACHO experiments and dynamical constraints, such as the unbinding of soft binaries (vulnerable to disruption from PBH encounters) or disk heating and instabilities caused by halo objects overheating stars in galaxtic discs or dragged by dynamical friction and similarly by the stability of tidal streams [1, 12, 13]. GW constraints are also important, as binaries would form by capture in mini-halos, and the LIGO limits may put stringent constraints on the fraction of dark matter in PBHs [6, 14, 15].

2 Seed Black Holes: Stars Light Up

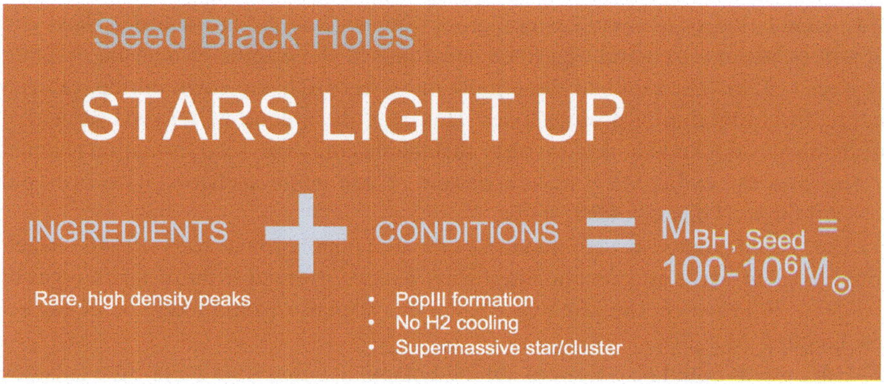

It is now well established that the properties of supermassive black holes (SMBH) found at the centers of galaxies today are tightly coupled to those of their hosts, implying a strong link between black hole and galaxy formation. The observation of luminous quasars at $z \approx 6 - 7$ implies that the first MBHs must have formed at very early times. The luminosities of these quasars are up to 10^{47} erg/s, which imply SMBH of a billion solar masses are already in place when the Universe is less than a billion years old. This concretely connects black hole formation and growth to the end of the cosmic dark ages when the first galaxies and structures light up the universe. We know that black holes are also fundamental constituents of the cosmic dawn.

Given the discovery of the first quasars at such early epochs, a seed black hole population much grow as the first stars and galaxies collapse. It is then a requirement that some of them will grow pretty much at their maximal rate since the time of formation (we will discuss this in the next section). It currently believed that the first 'seed' black holes must have appeared at early epochs, $z > 10 - 30$. Above we have discussed a putative population of primordial black holes. The origin and nature of this seed population remain uncertain due also to the lack of any conclusive observational evidence for such a population. Those primordial black holes were those forming at the epoch of radiation domination. More generally it is expected that other types of seed black hole should form after recombination. At this epoch the density fluctuations continue to grow and turn non-linear, this is what we refer to as structure formation. The collapse of these structures is halted when equilibrium is reached by virialization. It is within these dark matter halos that galaxies form from baryons cooling and collapsing. The first collapsing halos have virial temperatures smaller than $T \sim 10^4 K$. This is because at this temperature cooling from electronic excitation of atomic hydrogen becomes effective. In order for their gas to cool down and form the first stars, these halos must rely on the less effective H2 cooling (halos cooling via H2 are referred to as minihalos). The diagram in Fig. 4 shows an illustra-

tion of the seed black hole formation scenarios and associated conditions from the review by Regan and Hahnelt [16]. The fundamental scenarios for seed formation have been explored with state-of-the-art high resolution simulations of the collapse of isolated halos, or in small cosmological volumes over the last decade. These can all still be summarised and described according to the original schematic diagram by Rees (1978). The seeds of these most massive MBHs are sown during, or even before, the formation of protogalaxies. However, the precise MBH seed formation mechanism is not known, nor is it clear that there is only one seed formation channel at play over the entire MBH mass spectrum. Three major scenarios have been put forward for seed black holes (Fig. 5):
(1) They could emerge as the remnants of the first population of stars (PopIII) which could be as massive as a few hundred solar masses, formed from metal-free gas at $z \sim 20-30$ [17–19]. The remnant BHs in this scenario would likely be of mass a few hundred $M_\odot$ [20].
(2) Large seeds could also form by direct dynamical collapse in metal-free galaxies [21–24]. Under these conditions the BHs would be on the high mass end of the possible seed population, with masses of the order of $10^4-10^5\,M_\odot$.
(3) Conditions in the early universe could induce the formation of supermassive stars which also would quickly collapse into massive black holes. Alternatively, as in the original depiction by [25], the first episode of efficient star formation could lead to the formation of very compact nuclear star clusters [26, 27] where star collisions can then lead to the formation of a very massive star (VMS), possibly leaving a MBH remnant with mass in the range $\sim 10^2-10^4\,M_\odot$ [28–30].

We refer the reader to the comprehensive and recent review of [31] for more details of seed formation, and only highlight the most recent progress and challenges for this field here.

Population III remnants While this scenario seems the most natural for the formation of the first seeds, and it likely leads to the formation of the some of the first black holes it is not clear that this is the preferred path for the seeds of the massive black hole population or even more the first quasars. The main reason is that large uncertainties exist on the actual final mass of PopIII stars. It is not clear if the very first stars formed in singles or multiples per halo, and with a top-heavy mass function [26, 32]. Also, more recent simulations that are able to follow the evolution of the compact optically thick region further, have shown that even if a single object is initially formed from a $1000\,M_\odot$ clump of gas, it can collapse and fragment much further. This fragmentation can also be induced by the presence of external UV radiation and the temperature of the external cosmic microwave background floor (see the discussion [33]). We do not know if PopIII stars are indeed very massive [26, 34–36], and in particular if they are are able to reach the threshold of a few hundred solar masses for MBH formation (needed to form a BH rather than going through pair instability supernovae instead).

Gas-dynamical processes Via this route, massive black holes seeds are potentially formed. There are some pre-requisite conditions though for this possible channel, making it arguably a rare event that occurs in a small subset of early forming halos.

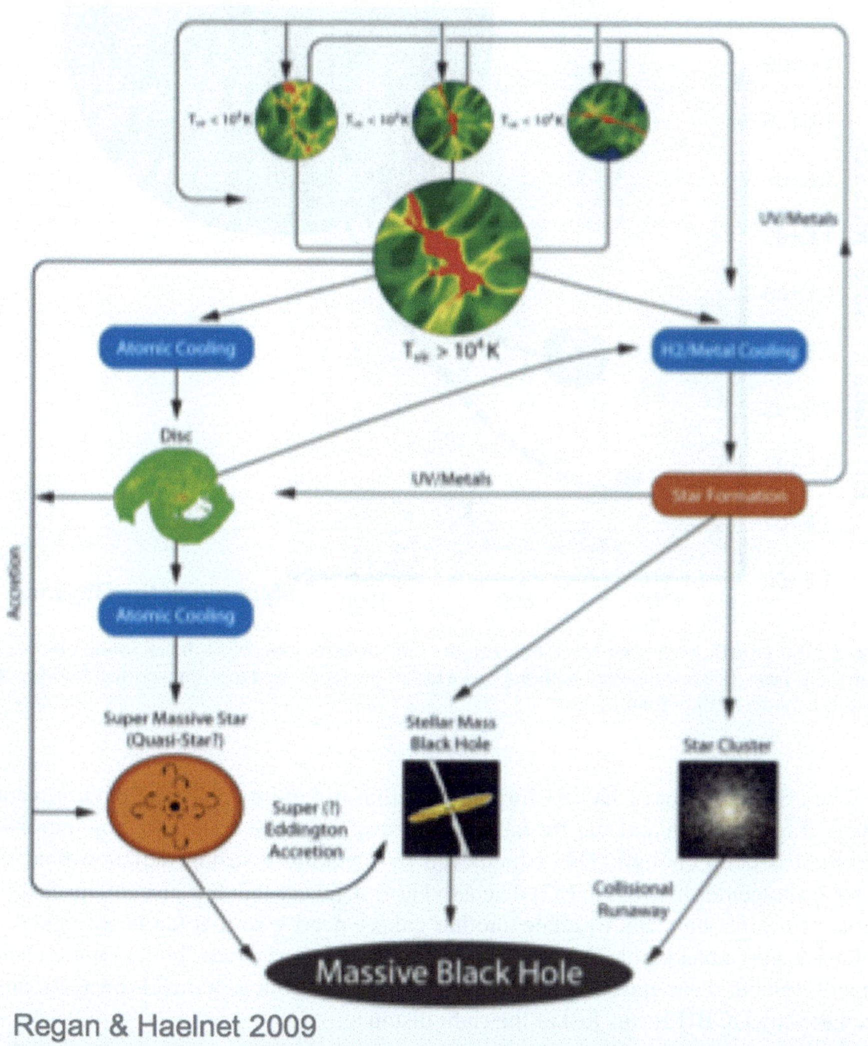

Fig. 4 Schematic for the formation scenarios for seed BHs, from [16]

A Lyman-Werner irradiating background needs to be present, so that the formation of molecular hydrogen (H2) in a primordial, star-less halo nearby is suppressed and the gas can only cool atomically. After reaching a virial temperature of $\sim 10^4 K$, these "atomic cooling" halos collapse isothermally to hydrogen number densities of 10^6 cm^{-3} without fragmenting. After this, the gas becomes optically thick to Ly-α radiation, so that it can begin a runaway collapse that eventually leads to the formation of a massive ($10^3 - 10^5$ M$_\odot$) direct collapse black hole (DCBH).

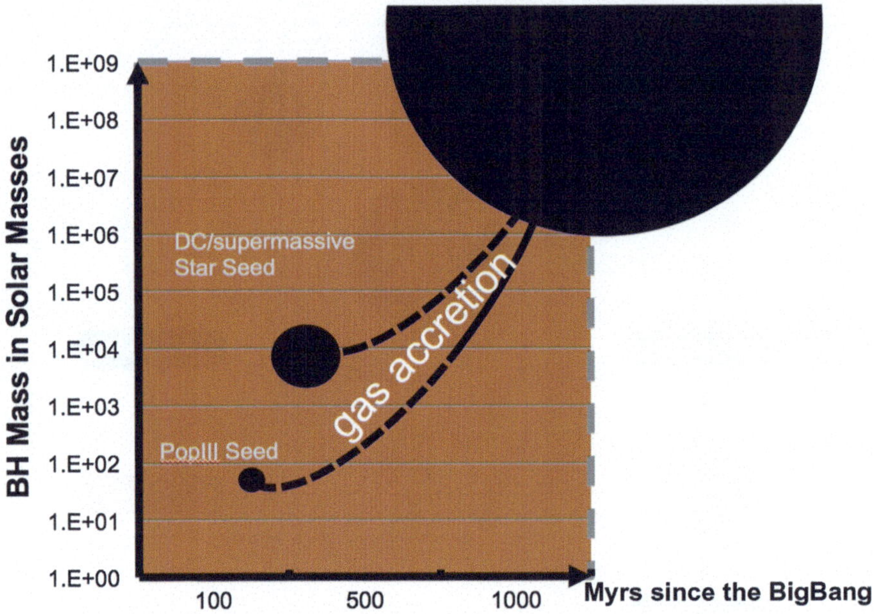

Fig. 5 The growth from seed black holes to the rare supermassive black holes in the observed high$-z$ quasars. In a few hundred million years a few of the seeds need to sustain critical Eddington growth to reach million solar masses

The crucial aspect of DCBH formation is that it requires a special environment where fragmentation and star formation must stay suppressed until the gravitational potential is deep enough. This is achieved by invoking a strong radiation field to destroy molecular hydrogen. [37] discussed how a galaxy that has just collapsed and started forming stars can irradiate another galaxy nearby so that the nearby galaxy collapses into a black hole. Massive stars in the first galaxy create both radiation and supernovae but the metals can lag behind the radiation front as it travels away. In this scenario the DCBH forms just at the right distance.

Some exciting new developments in simulations have however shown that the stringent requirement of the strong radiation field but no metals may be relaxed. [38] using the Renaisssance simulation, (see Fig. 6) have noticed that a small fraction of dark matter halos simply do not host any stars while still being massive enough to definitely be able to. The idea is that star formation can be suppressed if the halo grows fast enough (discussed also by Naoki Yoshida in the context of the formation of the first stars, and termed "dynamical heating").

Supermassive stars remnants Recent simulations have brought about another intriguing possibility for the formation of seed black holes. This is somewhat related to the formation of DCBH formation, but as in the recent work of Wise et al. discussed above allows relaxing the fine tuning requirements that the standard DCBH entail. In particular, [39] have performed cosmological hydrodynamical simulations

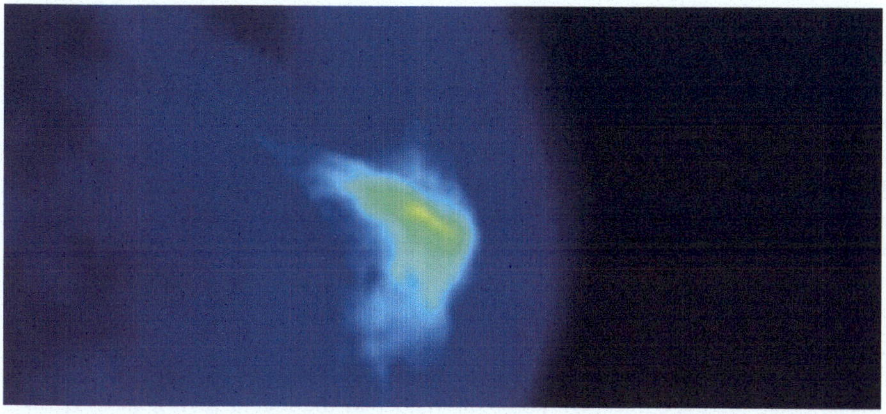

Fig. 6 One of the overdense regions containing a large amount of dark matter and creating a deep gravitational well in the Renaissance simulation. Gas falls into it and collapses into massive stars or directly into a black hole. Credit John Wise

Fig. 7 The gas density distribution around a newborn massive protostar. The left to right supersonic gas motion results in the non-spherical structure. Credit: Shingo Hirano

and shown that is possible to find regions in the initial density field in which supersonic gas motion prevents early gas cloud formation until a phase of extremely rapid halo formation takes places. In these regions (which were resimulated at high resolution using zoom in techniques, see Fig. 7) a protostar can form which grows by mass accretion up to 3–4 thousand solar masses. The massive star ends its life in a catastrophic collapse which leaves a black hole with roughly the same mass as the progenitor star [39, 40].

From these recent works it appears thus that heating provided by cosmological gas infall and mergers can play an important role, offering alternative routes to DCBHs. Another scenario, which also relies on heating from gas infall and accretion during the assembly of early galaxies, is the merger-driven model for DCHB presented originally in [41], and then later developed in [42, 43]. In this latter model there is no need of illumination by nearby star forming galaxies to dissociate molecular hydrogen, rather the heating is entirely provided by highly supersonic gas infall ad turbulence during major mergers between the most massive galaxies at $z \sim 8 - 10$.

Since the relevant cosmic epoch is significantly later than in the conventional models described above, gas is already, metal-enriched (solar metallicity or higher). Yet, because of the much larger mass that protogalaxies can have at this stage (the halo mass considered is around $10^{12}M_{\odot}$, which would correspond to the high end tail of the halo mass function at $z \sim 8$) the heating provided by supersonic quasi-radial gas infall in the much deeper potential well during the merger is much higher. Indeed in the published simulations it is found to be enough to suppress fragmentation in the core of the merger remnant. The gravitationally bound mass in such core is remarkable, in excess of $10^{9}M_{\odot}$. The physical conditions in the remnant are such that direct collapse into a supermassive black hole of a mass comparable to the bound gas core could occur via the general relativistic radial instability, a route named "dark collapse", which would naturally explain the rapid emergence of the bright high-z Quasars (see next section). We defer the reader to the recent review of [44] for a thorough discussion of all the proposed DCHB models, and of their relation with the physics of super-massive stars.

3 The First Massive Quasars and Galaxies: Cosmological Simulations

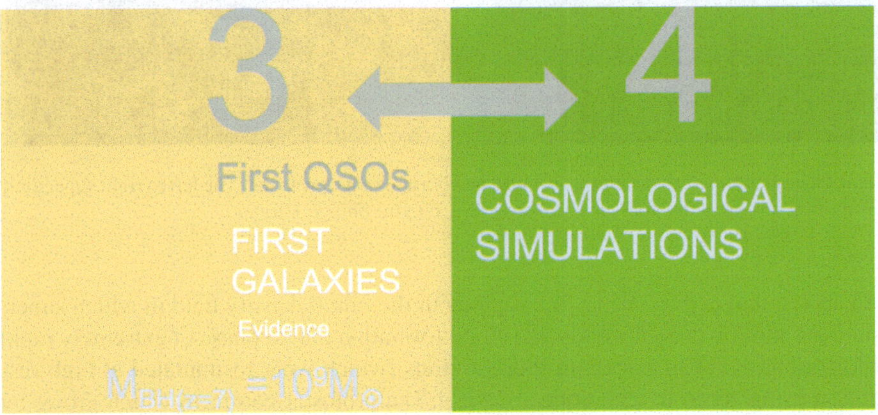

Quasars, powered by supermassive black holes, are the most luminous objects known and as a result they enable unparalleled studies of the Universe at the earliest cosmic epochs. The current record holder black hole with a mass of 800 million times that of the Sun has recently been discovered when the Universe was only 690 million years old—just five per cent of its current age [45], at the very dawn of galaxy formation. Now [46] presents a sample of a few tens of billion solar mass black holes have been discovered above $z \sim 6$ within the first billion years of cosmic history. It is one of the most challenging problems in galaxy formation and cosmology to explain how black holes as massive as those, (and as massive as those in the centers of today's galaxies) can grow in such a short time in the early Universe. As discussed in the previous section, the evidence for supermassive black holes within the first billion years puts into sharp focus that seed black holes must already be formed at early times, when the first stars collapse. Given any of these scenarios it is easy to work out that black hole seeds must grow at their critical Eddington rate for most of the time (this is the limiting rate when the outward continuum radiation pressure is equal to the inward gravitational force- equating these two quantities and solving for the luminosity gives the Eddington luminosity) For example, if the seed black hole is a few hundred solar masses then the seed needs to grow at Eddington for 100% of the time but if the seed is $10^5\,M_\odot$ then the requirement drops to 60% of time to reach a billion solar masses at $z = 6$. The question then is where and if/how the such high inflow rates can indeed be sustained for enough e-folds within our standard paradigm of structure formation. Understanding the conditions that make it possible for a few of the black holes to meet this condition means that extremely large volume simulation are needed to probe the highly biased, high density environments of the early Universe. This should guide one to identify the necessary conditions to explain the rare high-z quasars.

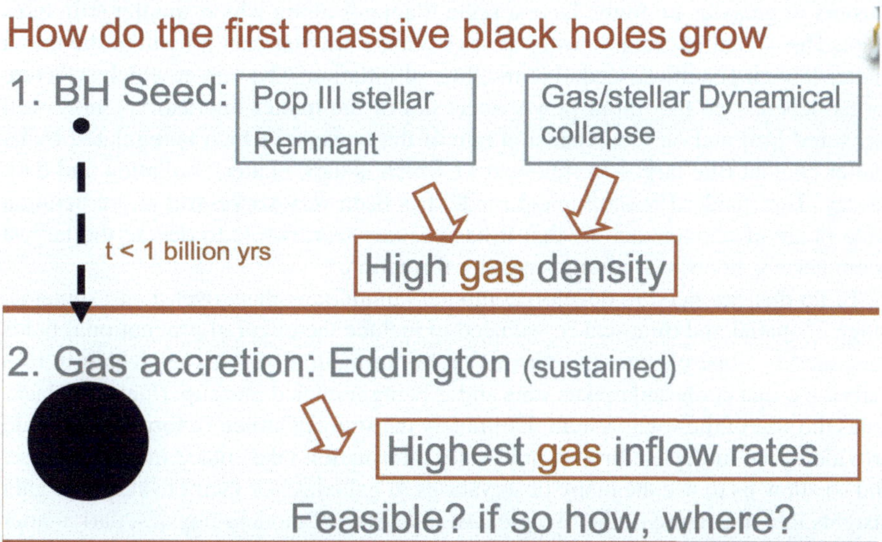

In our current understanding massive black holes form and reside at the centers of galaxies and hence they grow and merge closely intertwined with their host galaxies. The presence of luminous quasars observed within the first billion years of the Universe [31, 47] highlights the fact that the black hole seeds for the massive black hole population were assembled at the cosmic dawn, concurrently with the time of the formation of the first galaxies. In our standard Lambda Cold Dark Matter (ΛCDM) cosmology cosmic structure formation occurs hierarchically by the continuous merging of smaller structures and accretion of surrounding matter. SMBHs growth and evolution is expected to follow a similar process in which black hole seeds grow both though accretion and mergers with other BHs.

We will consider how the growth of seed black holes might occur in the context of our standard paradigm of structure formation. The bright quasars which are seen at the edge of the visible Universe are the rarest and most extreme objects one can imagine. Because of this extreme nature, quasars present cosmological theories with their toughest tests. In order to explain them, we must truly understand how structures form in the Universe. Tackling the formation of these first quasars numerically allows us to include the relevant complex physical processes.

In these lectures we will briefly introduce and review the progress in cosmological numerical simulations.

4 Cosmological Simulations

Structure formation and evolution in cosmology encompasses the description of the rich hierarchy of structures in our Universe, from individual galaxies and groups to clusters of galaxies up to the largest scale filaments along which smaller structures align. This so-called 'cosmic web' arises from the gravitational growth of the initial matter inhomogeneities, seeded at the time of inflation. The rate at which structure forms depends on the initial power spectrum of the matter fluctuations, now well measured [48] and on the expansion rate of the universe, which is regulated by its matter content (the largest component of which is dark matter), radiation and dark energy. The standard cosmological model has been very successful at predicting a wide range of phenomena, so that it has become worthwhile to devote the largest computer resources to studying structure formation.

To do this, we need to develop computer simulations that cover a vast dynamic range of spatial and time scales: we need to include the effect of gravitational fields generated by superclusters of galaxies on the formation of galaxies, which in turn harbor gas that cools and makes stars and is being funneled into supermassive black-holes the size of the solar system. Ultimately the study of structure formation should provide a true understanding of how galaxy formation takes place in the universe, and so allow us to use the many observations of galaxies and their clustering to gain insights into the nature of the two greatest mysteries of modern physics, dark matter and dark energy.

There are two conflicting requirements that make the study of hierarchical structure formation extremely challenging. In order to have a statistically significant representation of all structure in the Universe, the volume studied needs to be large but the individual particle mass needs to be small to adequately resolve the scale length of the structures which form and the appropriate physics. This implies a need for an extremely large N, where N is the number of particles. Depending on the problem, a dynamic range of 10^{10} or more can be necessary in principle.

The largest computer models of galaxy formation have traditionally involved the properties of dark matter only but the part of the Universe astronomers observe is made up of ordinary matter (gas, stars, etc.). In order to make direct contact with observations and predictions from our theories we must simulate the detailed hydrodynamics of the cosmic plasma. In addition, there is growing observational evidence for a close connection between the formation and evolution of galaxies and of their central supermassive black holes. Cosmic structure formation is nonlinear, involves a large variety of multi-scale physics and operates on large timescales making large scale numerical simulations the primary means for its study. Both approaches (dark matter only and full gas-dynamics, star formation and black hole physics) can be used in concert to make progress.

In hydrodynamic cosmological simulations, the complex non-linear interactions of gravity, hydrodynamics, forming stars, and black holes are treated in a large, representative volume of the universe. In this approach the physics at these much smaller galaxy scales is hence self-consistently coupled to large cosmological scales. These are therefore our most powerful predictive calculations linking the part of the universe we observe (stars, black holes, etc.) to the underlying dark matter and dark energy. They capture our most realistic understanding of black holes and their connection to galaxy formation.

Over the last few years it has become possible, with newly developed and more sophisticated codes, higher fidelity physical models as well as large enough computational facilities, to simulate statistically significant volumes of the universe (down to $z = 0$) with sufficient detail to resolve the internal structure of individual galaxies and follow the growth, mergers and evolution of black holes in their centers [49–52]. In Fig. 8 (left panel) we show a slice through a small piece of one simulation [53], where the black hole population can be seen to trace the overall large-scale structure. The richness of information available can be imagined, as we observe some of the millions of black holes simulated in our largest simulations. The prospect that we are in a position to use cosmology, i.e. the science of the Gigaparsec horizon, in our simulations to make predictions for the mass distribution in the inner regions of galaxies is extraordinary.

4.1 What We Simulate, Codes and Physics

To simulate structure formation in the Universe we need to account for its full cosmic matter-energy content. Matter comes in two basic types: ordinary *baryonic matter* (e.g. atoms, stars, planets, galaxies) which accounts for 15% of the total matter content, and *dark matter* which accounts for the remaining 85%. In addition, there is a mysterious *dark energy* field which actually dominates the energy density of the universe today, with a contribution of 75%, while matter constitutes only about 25%. The simulation code GADGET [54] and its various descendants is one of several hydrodynamic simulation codes adapted for use in cosmology. They use a number of different algorithms to self-consistently simulate all three components according to their appropriate physical laws. In this article we devote most space to GADGET, and its variant MP-GADGET [55], adapted for the largest simulations.

For dark matter, which is thought to behave as a perfectly collisionless fluid, the N-body method is used, where a finite set of particles samples the underlying

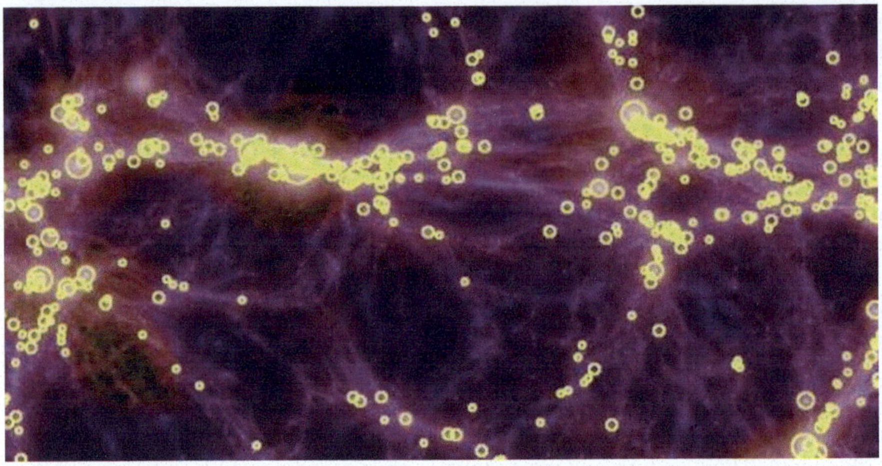

Fig. 8 The projected baryonic density field in a slice through a cosmological hydrodynamic simulation. The circles mark the positions of black holes with an circle area that scales with the BH mass

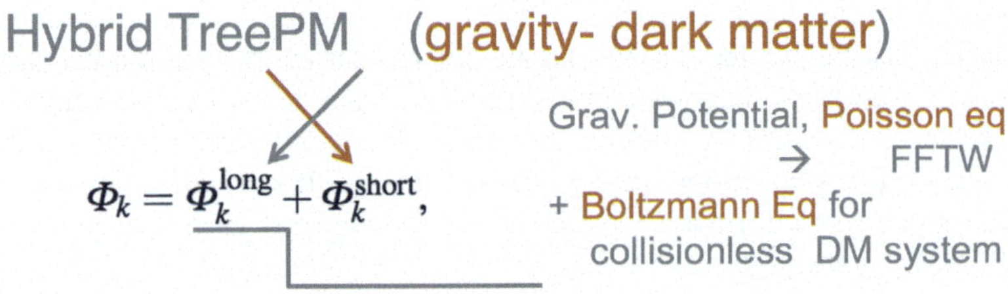

Fig. 9 The gravitational force calculation in many high resolution cosmological codes combines a Particle Mesh (PM) solver for long wavelengths and a direct sum over the nodes of a hierarchical tree for short wavelengths

distribution function. As the only appreciable interaction of dark matter is through gravity, and the evolution of the system obeys the Poisson-Vlasov equation. For the computation of the gravitational field, the code uses an FFT mesh solver on large-scales coupled to a hierarchical multipole expansion of the gravitational field based on a tree-algorithm [56] on small scales, leading to a uniformly high force resolution throughout the computational volume (Tree-PM algorithm, Fig. 9).

For the large scales, GADGET and many other cosmological codes use a hierarchical multipole expansion (organized in a "tree") to calculate gravitational forces. In this method, particles are hierarchically grouped, multipole moments are calculated for each node, and then the force on each particle is obtained by approximating the exact force with a sum over multipoles. The list of multipoles to be used is obtained with a so-called tree-walk, in which the allowed force error can be tuned in a flexible way. A great strength of the tree algorithm is the near insensitivity of its performance to clustering of matter, and its ability to adapt to arbitrary geometries of the particle distribution. While the high spatial accuracy of tree algorithms is ideal for the strongly clustered regime on small scales, there are actually faster methods to obtain the gravitational fields on large scales. In particular, the well-known particle-mesh (PM) approach based on Fourier techniques is probably the fastest method to calculate the gravitational field on a homogeneous mesh. The obvious limitation of this method is however that the force resolution cannot be better than the size of one mesh cell, and the latter cannot be made small enough to resolve all the scales of interest in cosmological simulations. Gadget therefore offers a compromise between the two methods. The gravitational field on large scales is calculated with a particle-mesh (PM) algorithm, while the short-range forces are delivered by the tree. Thanks to an explicit force-split in Fourier space, the matching of the forces can be made very accurate. With this TreePM hybrid scheme, the advantages of PM on large-scales are combined with the advantages of the tree on small scales, such that a very accurate and fast gravitational solver results. A significant speed-up relative to a plain tree code results because the tree-walk can now be restricted to a small region around the target particle as opposed to having to be carried out for the full volume.

SPH (Hydrodynamics – ideal fluid, baryons)

Cosmological expanding gas→ a is the scale factor

Euler
$$\frac{\partial \mathbf{v}}{\partial t} + \frac{1}{a}(\mathbf{v}\cdot\nabla)\mathbf{v} + \frac{\dot{a}}{a}\mathbf{v} = -\frac{1}{a\rho}\nabla P - \frac{1}{a}\nabla\Phi,$$

Continuity
$$\frac{\partial\rho}{\partial t} + \frac{3\dot{a}}{a}\rho + \frac{1}{a}\nabla\cdot(\rho\mathbf{v}) = 0$$

3rd law of thermodynamics
$$\frac{\partial}{\partial t}(\rho u) + \frac{1}{a}\mathbf{v}\cdot\nabla(\rho u) = -(\rho u + P)\left(\frac{1}{a}\nabla\cdot\mathbf{v} + 3\frac{\dot{a}}{a}\right)$$

Baryonic matter is evolved using a mass discretization of the Lagrangian equations of gas dynamics. In cosmology and galaxy formation simulations, both Eulerian and Lagrangian methods have been used to discretize the cosmic gas. Eulerian methods offer the principal advantage of high accuracy for shock capturing and low numerical viscosity. In MP-GADGET the baryonic matter is evolved using a mass discretization of the Lagrangian equations of gas dynamics. The code employs a particle-based approach to hydrodynamics, where fluid properties at a given point are estimated by local kernel-averaging over neighboring particles, and smoothed versions of the equations of hydrodynamics are solved for the evolution of the fluid (SPH). The hydrodynamics solver in MP-Gadget adopts the newer pressure-entropy formulation of smoothed particle hydrodynamics [57]. This formulation avoids non-physical surface tensions across density discontinuities.

With new hydrodynamic algorithms such as AREPO, [58] an unstructured Voronoi tessellation of the simulation volume allows for dynamic and adaptive spatial discretization, where a set of mesh generating points are moved along with the gas flow. This mesh is used to solve the equations of ideal hydrodynamics using a second order,

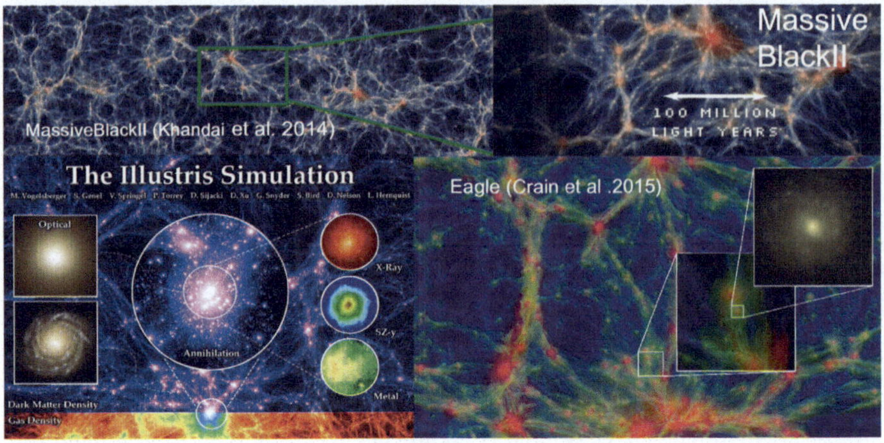

Fig. 10 A montage of images from the first set of simulations able to resolve formation galaxy formation in a uniform cosmological volume. Top: MassiveBlackII [51], bottom left, Illustris [65], bottom right, EAGLE, [49]

finite volume, directionally un-split Godunov-type scheme, with an exact Riemann solver. The code has been thoroughly tested and validated on a number of computational problems and small scale cosmological simulations [58–64] demonstrating excellent shock capturing properties, proper development of fluid instabilities, low numerical diffusivity and Galilean invariance, making it thus well posed to tackle the problem of galaxy formation. In recent years, these algorithms have enabled a computational approach to the full problem of galaxy formation (see Fig. 10).

Physical processes, galaxy formation: The galaxy formation model in both (a) and (b) suites is based on the inclusion of:
(i) Gas cooling and photo-ionization: the cooling function is calculated as a function of gas density, temperature, metallicity, UV radiation field, and AGN radiation field (in (b)).
(ii) Star formation and ISM model: the simulations adopt a subgrid model for the ISM, computing an effective equation of state assuming a two-phase medium of cold clouds embedded in a tenuous, hot phase. Star formation occurs stochastically and follows the Kennicutt-Schmidt law.
(iii) Stellar evolution and feedback: stellar populations return mass to the gas phase through stellar winds and supernovae. The simulations also employ a kinetic stellar feedback scheme, which generates a wind with velocity scaled to the local DM dispersion, and mass loading inferred from the available SN energy for energy-driven winds.

4.2 Black Holes in Cosmological Simulations

Here we review in some more detail the physical implementation of the BH physics in our cosmological simulations. We review the main elements of the subgrid model. The review here is not comprehensive and some variants and additional physical processes have been added in some of the recent numerical work in the field.

BH Seeds Since the SMBH seed formation processes (direct collapse, or stellar collapse, see [16] for review) are not resolved by cosmological simulations, it is assumed that every halo above a certain threshold mass hosts a central BH seed, so that BHs are always well resolved. Halos are selected for seeding by regularly running the "Friends-of-Friends" (FoF) halo finder on the dark matter distribution. The mass for the BH seed, $M_{BH,seed}$) and that of the halo, ($M_{FOF,seed}$) in each of the simulation runs is given in Table 1.

Black hole accretion BHs grow in mass by accreting gas from their environments according to a Bondi scaling until the accretion rate is at (or close to) the critical Eddington rate. There are some small differences however in exactly how the Bondi rate is calculated [66, 67]. For example, in the *Illustris* simulations, the relative velocity of black holes with respect to their surrounding gas is taken into account when estimating accretion. Also, a "pressure criterion" is introduced whereby the accretion rate estimate is lowered in cases where the gas pressure of the ambient medium cannot compress gas to a density exceeding the star-formation threshold in the vicinity of an accreting black hole. **BH Feedback** The feedback from SMBHs includes what is often referred to as quasar-mode or thermal feedback (this is in all simulations). A small fraction of the AGN bolometric luminosity is thermally coupled to the surrounding gas assuming a given radiative efficiency. The efficiency factor for the thermal feedback is $\epsilon_f = 0.05$ and thus effectively leads to an energy-driven outflow. *Illustris* also includes "radio" mode feedback which kicks in when the Eddington ratio drops below 0.05. A kinetic feedback is introduced in *IllustrisTNG* (no radio mode). In both Illustris and TNG consideration is given to the radiation field of AGNs, which heats surrounding halo gas, modifying its ionization state and net cooling rate. We provide the specific parameters related to the feedback models in Table 1.

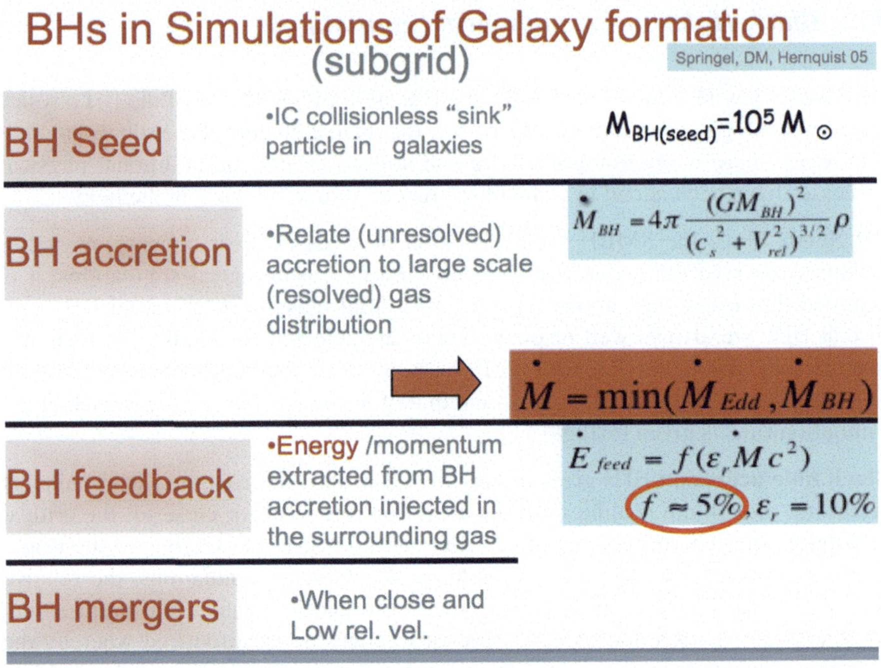

BH Feedback The feedback from SMBHs includes what is often referred to as a quasar-mode or thermal feedback (this is in all simulations). A small fraction of the AGN bolometric luminosity is thermally coupled to the surrounding gas assuming a given radiative efficiency. The efficiency factor for the thermal feedback is $\epsilon_f = 0.05$ thus effectively leads to an energy-driven outflow. *Illustris* also includes "radio" mode feedback which kicks in when the Eddington ratio drops below 0.05. A kinetic feedback is introduced in *IllustrisTNG* (no radio mode). In both Illustris and TNG consideration is given to the radiation field of AGNs, which heats surrounding halo gas, modifying its ionization state and net cooling rate.

BH mergers and repositioning Black holes merge when they approach the spatial resolution limit of the simulation. This means that a merger occurs when two BHs are separated by a distance that is smaller than both their respective smoothing lengths. Note that triple BH mergers can happen in a single time-step in the simulations, however this is extremely rare.

To prevent BHs from merging during fly-by encounters in simulations, a limit on the allowed relative velocity of the BHs is usually imposed. A repositioning scheme is applied to simulations whereby BHs are repositioned to the local gravitational potential minimum. This is to avoid BHs from spuriously wandering around due to two body scattering encounters with massive dark matter or star particles. This can occur as the simulations cannot model the dynamical friction acting on BHs when their masses are smaller than the typical mass resolution. Mass and spatial resolution requirements for all these processes are demanding (Fig. 11).

Table 1 Cosmological hydrodynamical simulations and BH model parameters

	Illustris	MBII	IllustrisTNG	BT
$L_{box}[h^{-1}Mpc]$ N_{part}	75 2×1792^3	100 2×1820	205 2×2500^3	400 (to $z = 7$) 2×7032^3
m_{DM} $[h^{-1}M_{\odot}]$	4.4×10^6	1.1×10^7	5.9×10^7	1.2×10^7
m_{gas} $[h^{-1}M_{\odot}]$	8.9×10^5	2.2×10^6	1.1×10^7	2.3×10^6
ϵ_{DM} $[h^{-1}kpc]$	1	1.85	2.1	1.5
ϵ_{gas} $[h^{-1}kpc]$	0.5	1.85	2.1	1.5
$M_{BH,seed}[h^{-1}M_{\odot}]$	$\mathbf{10^5}$	$\mathbf{5 \times 10^5}$	$\mathbf{8 \times 10^5}$	$\mathbf{5 \times 10^5}$
$M_{FOF,seed}[h^{-1}M_{\odot}]$	5×10^{10}	5×10^{10}	5×10^{10}	$\mathbf{5 \times 10^5}$
Accretion:	Bondi **Pressure crit.** Max Edd.	Bondi N/A Max **2*Edd.**	Bondi **Pressure crit.** Max Edd	Bondi N/A Max **2*Edd.**
Feedback:				
Radiative efficiency:	**0.2**	**0.1**	**0.2**	**0.1**
Thermal coupl. efficiency:	0.05	0.05	0.05	0.05
Radio feedback:	$\dot{M} < 0.05\dot{M}_{Edd}$	**N/A**	**N/A**	**N/A**
Radiative feedback:	Modify net cooling	**N/A**	Modify net cooling	**N/A**
Kinetic feedback:	**N/A**	**N/A**	Low accretion rates	**N/A**
Motion within halo:	Repositioned pot. min.	Repositioned pot. min.	Repositioned pot. min.	Repositioned pot. min.
Mergers:	W/I smoothing	W/I smoothing	W/I smoothing	W/I smoothing
	N/A	$\mathbf{v_{rel} < c_s}$	**N/A**	$\mathbf{v_{rel} < c_s}$

Simulation Validation Importantly, state-of-the-art simulations which include the full-physics of galaxy formation have been shown by and large to reproduce fundamental statistical properties and clustering of the galaxy population and of black holes/AGN. For example, the observed $M_{BH} - \sigma$ relation and total black hole mass density ρ_{BH}, as well as the quasar luminosity functions and its evolution in optical, soft and hard X-ray bands [51, 52, 68–70]. The simulations have advanced our understanding of a large number of galaxy formation, galaxy cluster and cosmology issues [52, 71–75] and also first galaxies and black holes at cosmic dawn [50, 55, 76–80].

This large body of already existing work, with a high level of ab-initio physics, makes these suites of simulations particularly attractive for studying the first quasars and growth of massive black holes in galaxies across the whole of cosmic history.

5 High Redshift Galaxies and Black Holes

How the first supermassive black holes form in the Universe and evolve into the bright quasars which are among the most distant objects seen is a question which has occupied much of the community. In order to address it, one must simulate a large

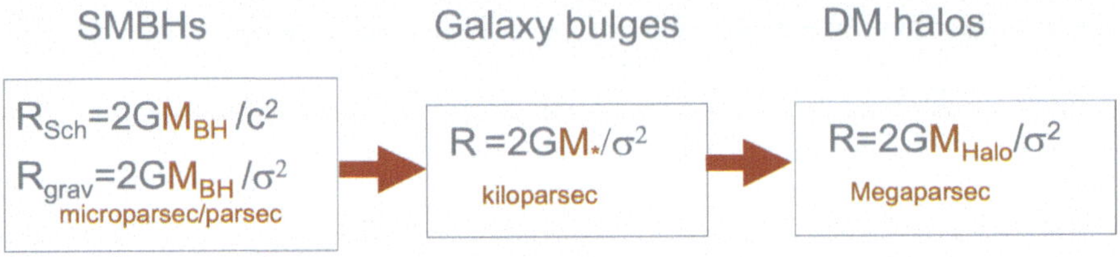

Fig. 11 Simulation of BH growth makes stringent demands of resolution, dictated by the BH-galaxy system

volume of the Universe, sufficient to include a statistical sample of bright quasars and their associated large scale structures, as well as having enough mass resolution to successfully model the physics and gas inflows close to black holes (the scales of an actual back hole accretion disk are still impossible to resolve in a fully cosmological simulations).

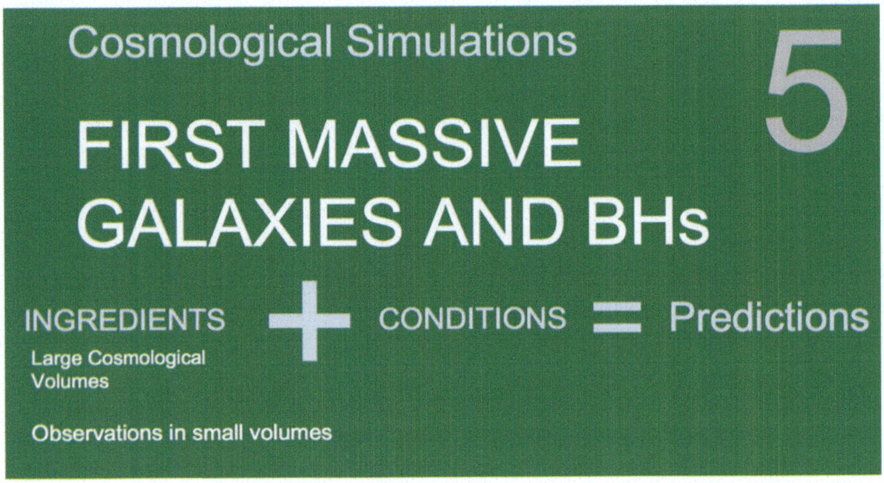

To decide which physical problem is best suited to which scale of computation one should consider both the space density of the relevant objects, which governs the size of the computational volume, and the mass resolution needed to follow the relevant small scale physical processes. In the case of the first quasars a volume of 1 Gpc/h is needed in order to have a statistical sample of at least 10 objects of the type seen at high redshifts by the Sloan Digital Sky Survey. The minimum mass resolution can be gauged using some smaller volume simulations that can be used to study convergence, i.e. carrying out cosmological simulations of black hole formation with different particle masses. At high redshifts the requirements are quite stringent, as galaxies are extremely compact, with sizes close to a few kpc in scale at best, and the structure and inflow within each galaxy needs to be resolved. This pushes the resolution requirements to only a few hundred parsec.

As simulating these processes can now be done in the petascale regime with High Performance Computing (HPC) systems, it has finally become possible with the newly developed MP-GADGET cosmological code to answer questions in cosmology which require simulations of the entire visible universe at high mass and spatial resolution. This applies to making direct predictions for what should be seen in future observations and surveys as well as using the simulations as tools to make mock future surveys. The simulations follow the evolution of the matter, energy and radiation in model universes from the Big Bang to the present day. For these projects, the actions of gravity, hydrodynamics, forming stars, black holes, molecular gas, inhomogeneous ionizing radiation and more are all included.

With state-of-the-art current HPC Facilities, for example, BlueWaters at NCSA in Illinois, USA, we have reached the point where the required number of particles (about one trillion) can be contained in memory, and the petaflop computing power is available to evolve them forward in time. On BlueWaters therefore is has become possible to make this qualitative advance, running, arguably, the first complete simulation (at least in terms of the hydrodynamics and gravitational physics) of the creation of the first galaxies and large-scale structures in the universe. This is the BlueTides simulation, which we use to discuss the emergence of the first quasars.

5.1 The Bluetides Simulation

The development of cosmological codes such as MP-GADGET has now been adapted to Petascale supercomputers and these resources have been used to understand how supermassive blackholes and galaxies formed, from the smallest to the rarest and most luminous. With nearly one trillion particle we have carried out the BlueTides simulation [50, 55, 79, 81] on BlueWaters. BlueTides was a full-machine run on Blue Waters, using the full Petaflop of memory available. The simulation aims to understand the formation of the first quasars and galaxies, and these role of these processes in the reionization of the Universe (Fig. 12).

BlueTides is the largest simulation yet run with full physics (hydrodynamics, star formation, black holes), and is targeted at the early Universe of galaxies and quasars. The simulation can be compared to cutting edge observations from the Hubble Space telescope, finding good agreement with the properties of observed galaxies when the Universe was only 5 percent of its current age ($z = 8$). This simulation actually covers a sky area 300 times larger than the largest current survey with the HST, allowing us to make predictions for what the upcoming successors to Hubble will see.

In the coming decade, a new generation of astronomical instruments, all in the billion dollar class will start making observations of the Universe during the period of the first stars and quasars, and opening up the last frontier in astronomy and cosmology. Those that are specifically targeting this epoch as their highest priority include the Square Kilometer Array radio telescope, the NASA James Webb Space Telescope, the successor to Hubble, and several huge ground-based telescopes, such as the Thirty Meter Telescope, the European Extremely Large Telescope, and the Giant Segmented Meter Telescope, each of which have collecting areas an order of magnitude larger than the current largest telecopes. The scientific community has obviously decided that research targeting the epoch of the first quasars and black hole formation and growth matters enormously. These observation and experiments will gain a lot of value if we have theories to test our models. It is crucial therefore to build equally powerful theoretical tools. The combination of Petascale HPC resources and the newest simulation codes are the equivalent to billion dollar-class theory programs in cosmology. These large simulations are important because without them there are no reliable ways to know what the cold dark matter cosmology predicts for the first stars and galaxies and their properties. The best, most complete and also largest simulations must be carried out in concert with these ambitious upcoming observational programs. This is what BlueTides aims to be.

The first quasars: high density and low tidal fields We have investigated in detail what causes the growth of the most massive black holes in the early Universe, and found that tidal fields in the large-scale (megaparsec and more) environments of black hole hosting galaxies play a critical role. The necessary growth is consistent with the detection of highly luminous quasars at $z > 6$, and implies sustained, critical accretion of material to grow and power them. Given a black hole seed scenario, it

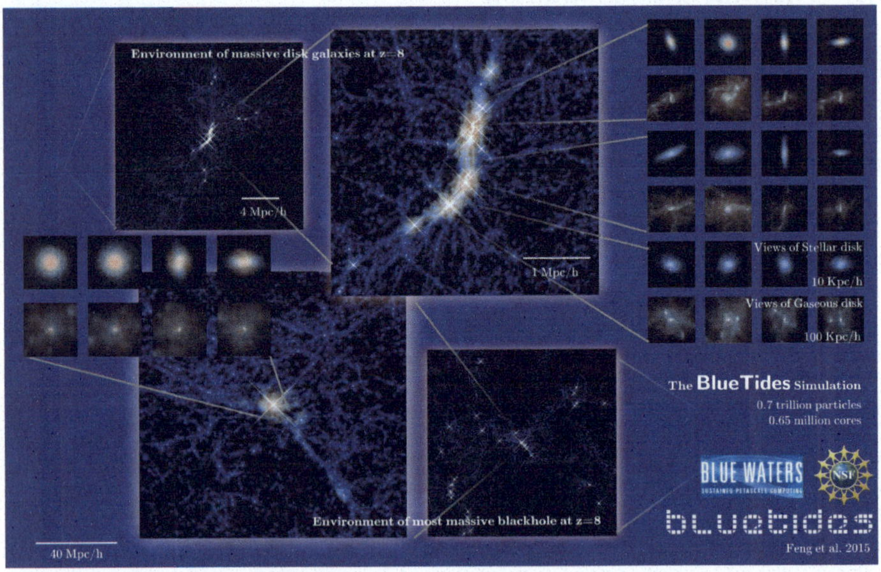

Fig. 12 An overview of the BlueTides simulation at redshift $z = 8$. The background shows a slice through the entire volume, and the bottom insets show the environment of the most massive black hole (bottom right), and its host galaxy in different projections (bottom left). The top insets show the host halo of some of the most massive disk galaxies (left) and their projected stellar and gas distributions (right)

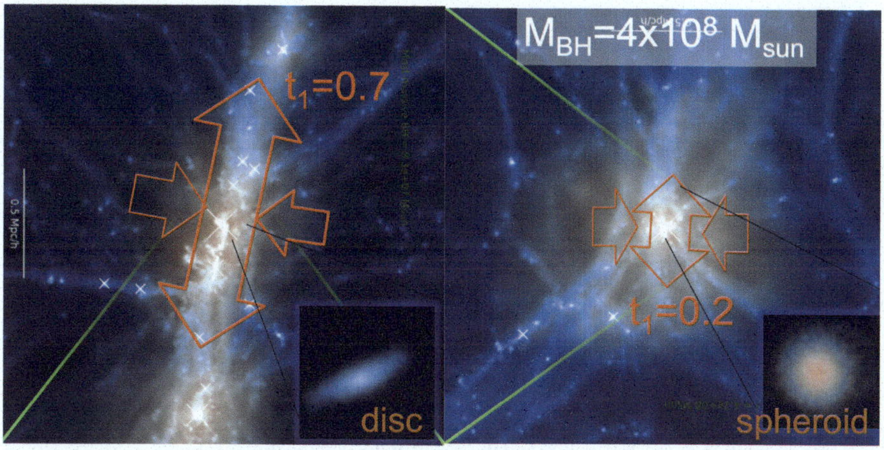

Fig. 13 The local environment of two galaxies in the BlueTides simulation at $z = 8$. The left panel shows a disk galaxy in a high tidal field environment (t_1 component of tidal tensor equal to 0.7), shown schematically with arrows. The right panel shows a spheroidal galaxy hosting the most massive black hole. It is in a region with a lower tidal field ($t_1 = 0.2$) [50]

is still uncertain which conditions in the early Universe allow the fastest black hole growth. Large scale hydrodynamical cosmological simulations of structure formation allow us to explore the conditions conducive to the growth of the earliest supermassive black holes. We use the cosmological hydrodynamic simulation BlueTides, which incorporates a variety of baryon physics in a $(400\,\mathrm{Mpc/h})^3$ volume with 0.7 trillion particles to follow the earliest phases of black hole critical growth.

At $z = 8$ the most massive black holes (a handful) approach masses of 10^8 Msun with the most massive (with $M_B H = 4 \times 10^8\,\mathrm{M}_\odot$ [50]) being found in an extremely compact spheroid-dominated host galaxy. Examining the large-scale environment of hosts, we find that the initial tidal field is more important than overdensity in setting the conditions for early BH growth. In regions of low tidal fields gas accretes 'cold' onto the black hole and falls along thin, radial filaments straight into the center forming the most compact galaxies and most massive black holes at earliest times. Regions of high tidal fields instead induce larger coherent angular momenta and influence the formation of the first population of massive compact disks. This can be seen clearly in Fig. 13 where the environment of the most massive black hole is visualized alongside a massive disk galaxy [50, 79].

The massive BlueTides dataset enables statistical measures to made of the impact of tidal field on black hole host galaxy morphology as well as black hole accretion rate, to show that the observed trends are significant. This can be seen in Fig. 14.

The extreme early growth depends on the early interplay of high gas densities and the tidal field that shapes the mode of accretion. Mergers play a minor role in the formation of the first generation, rare massive BHs. BlueTides shows that these conditions arise naturally and at the right number density in the standard cosmological

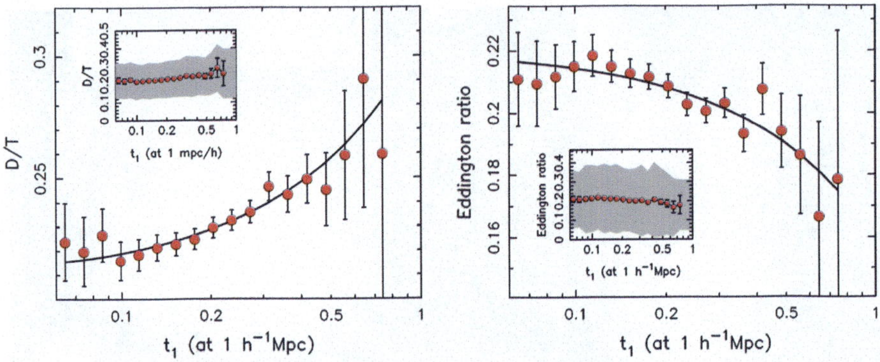

Fig. 14 The effect of large scale tidal field around galaxies on the disk to total ratio (left panel) and Eddington ratio of accretion onto their central black holes (right panel). Results are from the BlueTides simulation at redshift $z = 8$, and the x-axis indicated the dimensionless t_1 component of the tidal tensor smoothed on a 1 Mpc/h scale

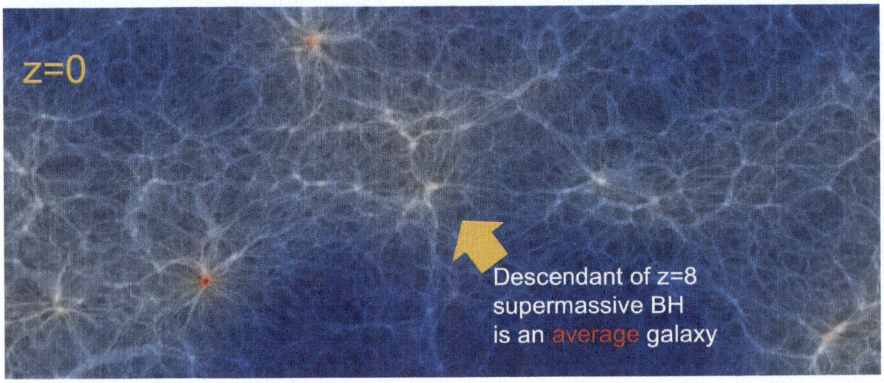

Fig. 15 The redshift $z = 0$ environment of what was the most massive black hole in the BlueTides simulation at redshift $z = 8$. The BTMassTracer simulation, a dark matter only simulation was used to evolve the density distribution of BlueTides to $z = 0$

model, as long as the resolution and box size are sufficient to allow proper simulation. This combination had not been achievable until recently [50, 79].

Where are the first quasars today? With the BTMassTracer simulation, a dark matter only realization of the Bluetides simulation, we are able to rapidly trace what would happen to the black holes in the full BlueTides run if it was run to the present day. We do this by following the particles positioned at the potential well minimum of each galaxy using a non-hydrodynamic simulation which is run to redshift zero. In Fig. 15 we can see that the descendant of the most massive galaxy at redshift $z = 8$ is not an exceptional galaxy halo. The first, most massive quasars at high redshift are not the most massive clusters in todays' Universe. This can also be seen in the

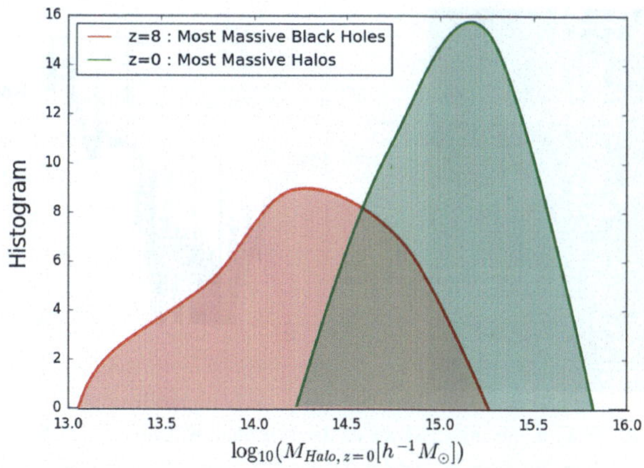

Fig. 16 Histograms showing that the descendants of halos hosting the most massive black holes at high redshift ($z = 8$) are group size, and not cluster sized at $z = 0$. Results from the BlueTides and BTMasstracer simulations [79]

histograms (Fig. 16) of the halo mass distribution of the descendants of the top 20 most massive black holes (red) and the top 20 most massive halos at $z = 0$ [79].

The first giant black holes live in massive but tiny galaxies The most distant known quasar recently discovered by [45] is seen 690 Myr after the Big Bang, at the dawn of galaxy formation. We explore the host galaxy of the brightest quasar in the large volume cosmological hydrodynamic simulation BlueTides, which in Phase II has reached these redshifts. The brightest quasar in BlueTides has a luminosity of 10^{13} times that of the sun and a black hole mass of 700 million solar masses, comparable to the observed quasar (Fig. 17). The quasar resides in a rare halo of mass 10^{12} solar masses, this is a host galaxy of Milky Way mass (See Fig. 18 for its large-scale environment) [78, 79].

We derive quasar and galaxy spectral energy distributions (SEDs) in the mid and near infrared and make predictions in JWST (James Webb Space Telescope, successor to Hubble) bands. We predict that a significant amount of dust is present in the galaxy. We present mock JWST images of the galaxy (Fig. 19). The host galaxy is detectable in NIRCam filters, but it is extremely compact (10 times smaller than our Milky way). JWST's exquisite sensitivity, resolution and wide wavelength coverage will be necessary to to do this (it would be undetectable with the current Hubble Space Telescope) [79].

Do the first supermassive black holes ever stop growing: AGN feedback at high-z? Many theoretical models predict that quasar driven outflows account for the observed quenching of star formation in massive galaxies. There is growing observational evidence for quasar-launched massive outflows even in the very early universe—this is referred to as quasar feedback. We have studied the feedback around the highest redshift quasar in the BlueTides simulation, the largest volume cosmological hydrodynamic simulation so far carried out. We have made predictions for

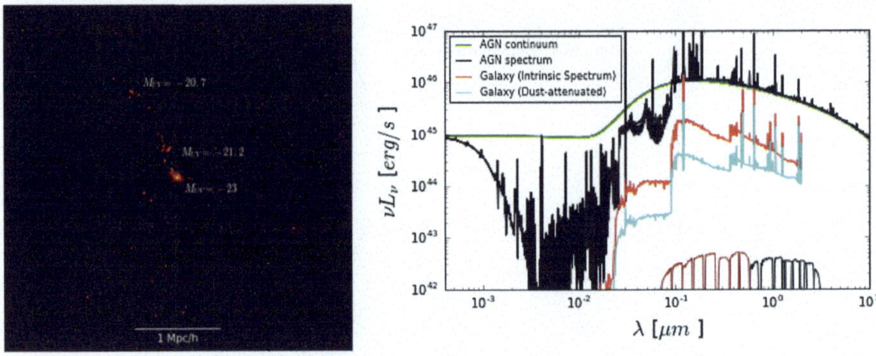

Fig. 17 A full-field view of the host galaxy of the brightest redshift $z = 7.5$ quasar in the BlueTides simulation. The left panel shows the JWST field of view. Two companion galaxies would be detectable by JWST. The right panel shows the galaxy and quasar spectra. The intrinsic UV magnitude of the galaxy is –23.1, which is roughly 2.7 magnitudes fainter than the quasar's magnitude of –25.9 [79]

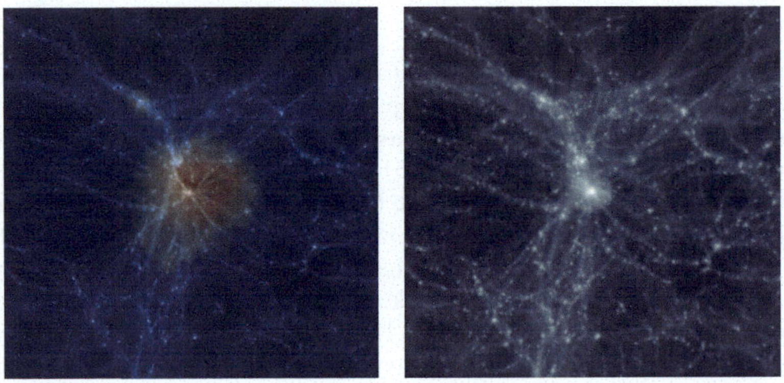

Fig. 18 Images showing the distribution of gas and dark matter in the region centered at the most massive black hole of size 4h-1Mpc corresponding to the JWST field of view at $z = 7.5$. Left: Distribution of gas where the intensity of the blue region represents the density and the color scale represents the temperature of the gas. Right: the corresponding dark matter density [79]

gas outflows around the brightest known quasar which hosts the most massive black hole in the simulation volume, consistent with the current record holder for high-z quasars. We predict that there are significant outflows around this quasar (Fig. 20). The gas is blown out from the galaxy, the quasar stops growing and so does its host galaxy. The outflow gas contains a cold, dense molecular component that originates from the inner region of the halo, within a few kpc of the central black hole. This would be observable in CO emission at radio wavelengths [78] (Fig. 21).

The velocities of the outflow gas reach thousands of km/s, within which the molecular component has mass averaged outward radial velocity of 1300 km/s, consistent with observations. The averaged outflow rate has an enormous value, about 200–300

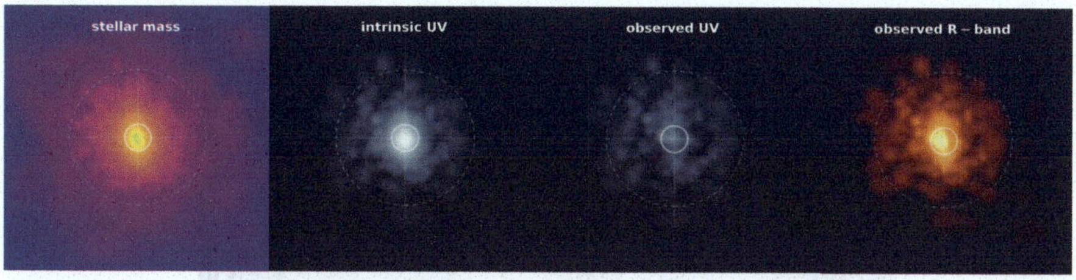

Fig. 19 Close up images of the host galaxy of brightest quasar in BlueTides at $z = 7.5$ in a region of size 10 kpc in physical units with a resolution of 0.1 kpc. The inner circle is of radius 1 kpc (0.2 arcsec)) and the outer circle depicts a radius of 5 kpc (1 arcsec). From Left to Right, the images show the distribution of stellar mass, intrinsic UV-band luminosity, observed UV-band luminosity and observed R-band luminosity [79]

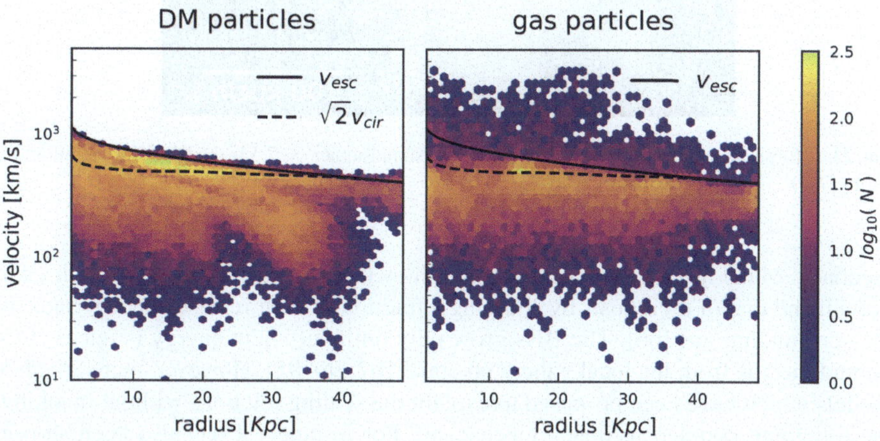

Fig. 20 Outflow gas in and around the halo of the brightest quasar at $z = 7.5$ in the BlueTides simulation. We show the velocity distribution of dark matter (left panel) and gas particles (right panel) at $z = 7.54$. The escape velocity profile (black line) aligns with dark matter particles well. The gas particles whose velocity exceed the escape velocity are outflowing

solar masses/yr, or one hundred times greater than the current outflow from our own Milky Way galaxy. We predict that the outflows we have seen in the simulated galaxy halo are likely to be present in the observed quasar. In addition, the presence of such significant quasar driven outflows may help explain the low star formation rate in the host galaxy and its stunted growth while allowing significant metal enrichment over the scale of the halo and beyond.

The Black-Hole Galaxy Relations at high redshifts Over the last three decades, scaling relations between the mass of supermassive black holes (SMBHs) and several stellar properties of their host galaxies such as bulge stellar mass and bulge velocity dispersion [82–89] have been discovered and measured for galaxies with black holes (BHs) and active galactic nuclei (AGN) from $z = 0$ and up to $z \sim 2$ (using different techniques). A popular way to interpret the scaling relations is by invoking AGN

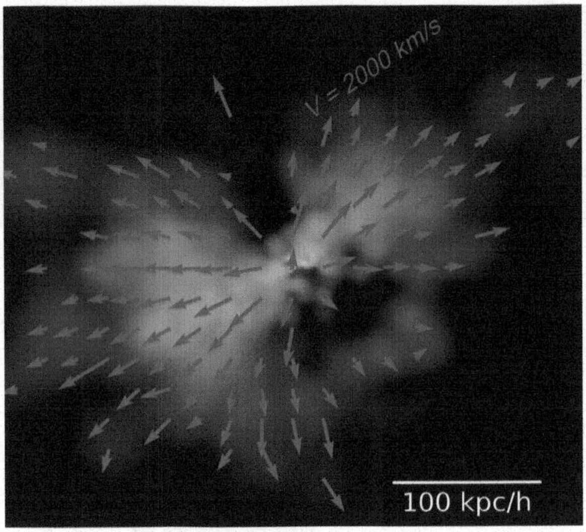

Fig. 21 Morphology of outflow gas from the brightest quasar at $z = 7.5$ in the BlueTides simulation (from [78])

feedback. Many models (and simulations) show that the SMBHs regulate their own growth and that of their hosts by coupling a fraction of their released energy back to the surrounding gas [66]. The BHs grow only until sufficient energy is released to unbind the gas from the local galaxy potential [67, 90–95]. However, there are also models which have been proposed to explain the scaling relations without invoking the foregoing coupled feedback mechanism. For instance, it has also been shown that dry mergers can potentially drive BHs and their hosts towards a mean relation [96, 97].

Regardless, studying the scaling relations in both observations and simulations is essential for understanding the coupled growth of galaxies and BHs across cosmic history [77].

An important related question is when the scaling relations are established, and if they still persist at higher redshifts when the first massive BHs form ($z > 6$). Many theoretical models have been developed to understand the origin of these relations. Several cosmological simulations that follow the formation, growth of BHs and their host galaxies have successfully reproduced the scaling relations at low-z (which we will review later) leading to an overall agreement of the scaling relations from observations and simulations at low-z, linking the growth of SMBHs to the growth of their hosts via AGN feedback.

An important related question is when the scaling relations are established, and if they still persist at higher redshifts when the first massive BHs form ($z > 6$). Bluetides, which has a large population of black holes and galaxies across a large range of masses which can be used to understand the assembly of the galaxy-black hole relations at high-z.

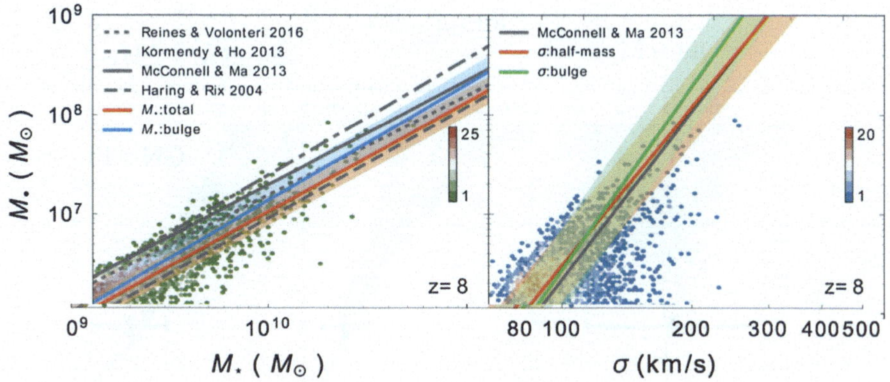

Fig. 22 High redshift ($z = 8$) scaling relations between BH mass and galaxy stellar mass (left panel) and stellar bulge velocity dispersion (right panel). The points shown are from the BlueTides simulation and the lines show different fits to low redshift observational results [77]

For each galaxy in the simulations we know the mass of its supermassive black hole ($M_\bullet$) and can measure host galaxy properties (stellar mass, $M_\star$, and velocity dispersion, σ). With these we can check if there is a link between the growth of black holes (BHs) and that of their hosts and investigate if and how the BH-galaxy relations are established in the high-z universe. Results are shown in Fig. 22. We find [77] the following $M_\bullet - M_\star$ and $M_\bullet - \sigma$ relations at $z = 8$: $\log_{10}(M_\bullet) = 8.25 + 1.10\ \log_{10}(M_\star/10^{11} M_\odot)$ and $\log_{10}(M_\bullet) = 8.35 + 5.31\ \log_{10}(\sigma/200kms^{-1})$ at $z = 8$, both fully consistent with the local measurements. The slope of the $M_\bullet - \sigma$ relation is slightly steeper for high star formation rate and $M_\star$ galaxies while it remains unchanged as a function of Eddington accretion rate onto the BH. The intrinsic scatter in $M_\bullet - \sigma$ relation in all cases ($\epsilon = 0.36$) is larger at these redshifts than inferred from observations and larger than in $M_\bullet - M_\star$ relation ($\epsilon = 0.14$). We find the gas-to-stellar ratio $f = M_{\rm gas}/M_\star$ in the host (which can be very high at these redshifts) to have the most significant impact in setting the intrinsic scatter of $M_\bullet - \sigma$. The scatter is significantly reduced when galaxies with high gas fractions ($\epsilon = 0.28$ as $f < 10$) are excluded (making the sample more comparable to low-z galaxies); these systems have the largest star formation rates and black hole accretion rates, indicating that these fast-growing systems are still moving toward the relation at these high redshifts. These kind of highly star forming galaxies, with high gas fractions do not have local counterparts indicating that the increase in scatter is not in tension with the local relations. Such high gas fraction systems have the largest star formation rates and black hole accretion rates, indicating that these fast growing systems at high-z are in the process of reaching the relation [77].

The brightest, highest redshift galaxy in the Universe Using deep Hubble and Spitzer observations, [98] have identified a bright star forming galaxy candidate at $z = 11$, GN-z11. The presence of GN-z11 implies a number density approaching 10^{-6} Mpc^{-3} roughly an order of magnitude higher than the expected value based

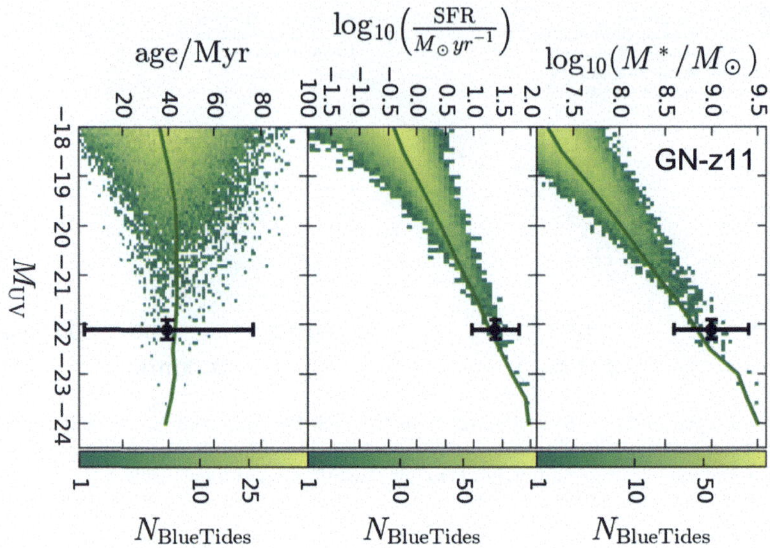

Fig. 23 The age, star formation rate and stellar mass of the highest redshift observed galaxy GN-z11 [98] compared to galaxies in the BlueTides simulation at $z = 11$

on extrapolations from lower redshift. Using the unprecedented volume and high resolution of the bluetides cosmological hydrodynamical simulation we study the population of luminous rare objects at $z > 10$. The luminosity function in Bluetides also implies an enhanced number of massive galaxies, consistent with the observation of GN-z11. We find about 30 galaxies with $M < -22$ at $z = 11$ in the Bluetides volume, including a few objects about 1.5 magnitudes brighter. The probability of observing GN-z11 in the volume probed by Oesch et al. is 20%. The predicted properties (star formation rates, stellar masses, and age) of the rare bright galaxies at $z = 11$ in BlueTides closely match those inferred from the observations of GN-z11. Bluetides predicts a negligible contribution from faint AGN in the observed SED. At these epochs, the brightest galaxies have black holes with a black hole mass close to a million solar masses, typically providing less than 20% of the total UV luminosity [99, 100] (Fig. 23).

In BlueTides, we have also found that the first massive galaxies to form are predicted to have extensive rotationally supported disks. Although their morphology resembles in some ways Milky Way types seen at much lower redshifts, the high-redshift galaxies we have found are smaller, denser, and richer in gas than their low-redshift counterparts. Images of a sample of disks in BlueTides are shown in Fig. 24 [55].

Milky-Way mass, compact disks at $z > 6$ From BlueTides, one can 'detect' galaxies from mock observations. We have applied observational selection algorithms (the widely used SourceExtractor software) to the simulated sky maps and created catalogs of millions of galaxies relevant for the upcoming frontier fields (JWST and

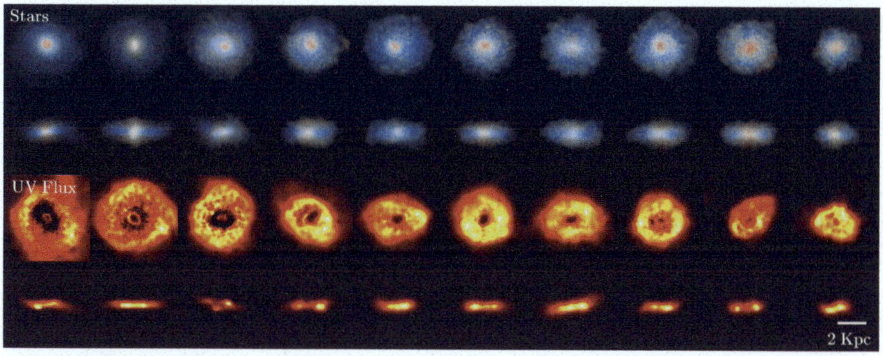

Fig. 24 Images of some representative massive disk galaxies (halo masses $\sim 10^{12} M_{\odot}$) from the BlueTides simulation at $z = 8$. We show face and edge on views of the stellar density and the UV rest frame flux

WFIRST). Making use of the high resolution of BlueTides we have made detailed images of individual galaxies, uncovering the striking and unexpected population of large Milk Way-mass disk galaxies (see Fig. 24) present when the universe was 5% of its present age to be observed in the near future [55].

In particular, from a kinematic analysis of a statistical sample of 216 of the most massive galaxies at redshift $z = 7 - 10$ in Bluetides we have found that disk galaxies make up 70% of the population of galaxies of roughly Milky Way mass or greater. Cold dark matter cosmology therefore makes specific predictions for the population of large galaxies 500 million years after the Big Bang. The sizes of these galaxies compared to current observations are shown in Fig. 25. It can be seen that current observations probe much fainter galaxies as no existing space telescopes have a wide enough field of view to capture the bright and rare galaxies that have formed in BlueTides. We therefore argue that widefield satellite telescopes (e.g., WFIRST) will in the relatively near future discover these first massive disk galaxies. This is an exciting development, as it implies that large galaxies are present at the end of the Dark Ages. The simplicity of their structure and formation history should make new tests of cosmology possible.

AGN and Reionization Bluetides captures the growth of BHs that make up the lower luminosity AGN population, beyond what is currently observed. This is helpful for our understanding of the AGN contribution to their host galaxies evolution and in general to the process of reionization. It is well known that by $z \sim 6$ the ionizing radiation emitted by bright Sloan-like quasars alone is insufficient to reionize the IGM. However, at $z > 5$, the extrapolation of the high-z QSO LFs and their redshift evolution as well as the space density of lower luminosity AGN is likely crucial for understanding their contribution to reionization. The long standing hypothesis that reionization is due to Lyman Break Galaxies alone requires a significant contribution of faint dwarf galaxies and a Lyman continuum photon escape fraction of the order of ~20%, which has now been brought into tension with present observational

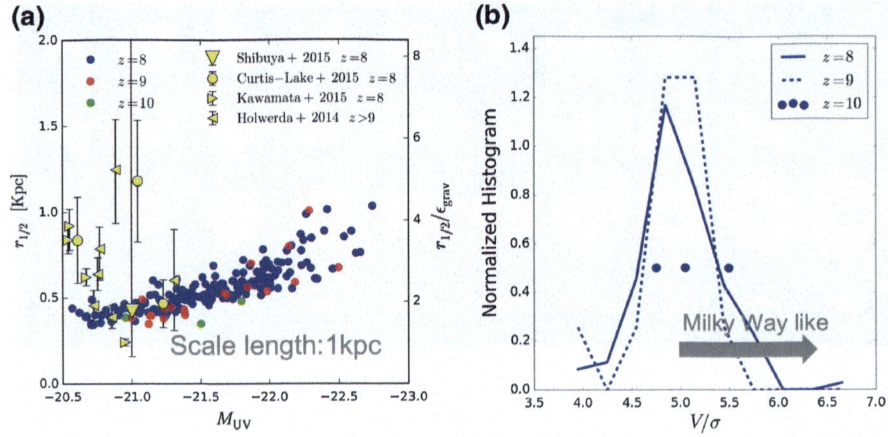

Fig. 25 Galaxy properties at high redshift in the BlueTides simulation. Panel **a** shows sizes (scale radii) compared to observations. Panel **b** show histograms of the kinematic diagnostic quantity V/σ, the ratio of circular velocity to velocity dispersion. Milk Way-like disks lie on the right hand side

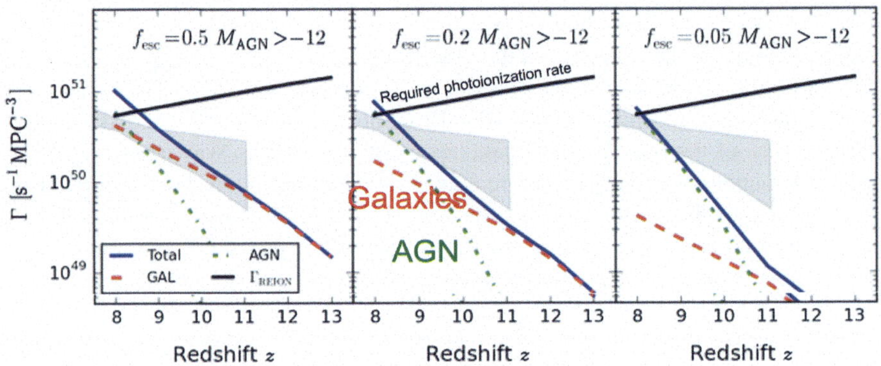

Fig. 26 Reionization constraints from the BlueTides simulation. We show the photoionization rate as a function of redshift for three different values of the photon escape fraction from galaxies. The required photoionization rate for reionization is shown as a black solid line in each panel. Galaxies can reionize the universe for high escape photon fractions but AGNs can contribute significantly [81]

constraints Clearly, examining the combined contribution of faint AGN and galaxies is crucial. With our simulations we are able to tackle the outstanding questions of whether AGNs or galaxies are responsible for reionization, and how many galaxies host an AGN at high-z [81] (Fig. 26).

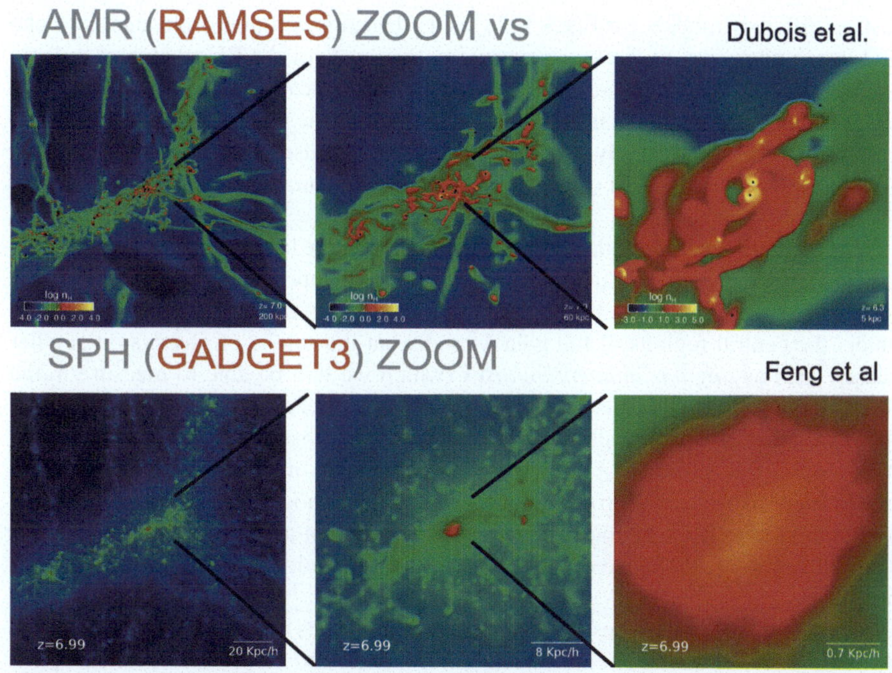

Fig. 27 Examples of zoomed simulations, resimulating regions chosen from the MassiveBlackI simulation with higher resolution (with Gadget) and with higher resolution and a different code (the Adaptive Mesh Refinement code RAMSES). The growth of the central black hole in the zoomed re-simulations is very similar

5.2 Zooming in: High-z Massive Galaxies and Black Holes

With current high-performance (Petascale) computers and the aggressive development of cosmological codes, we have just entered the regime in which hydrodynamical cosmological simulations of sizeable fractions of the entire visible universe are carried out at exquisite mass and spatial resolution. We are hence able to build mock quasar/galaxy datasets that provide the statistical power to interpret observations and make predictions for future surveys. Bluetides is an example of such a simulation, targeted at the high-z regime and making direct contact with the observations of the first quasars.

Two main classes of simulation are standardly employed for such studies. A complementary approach to large, uniform volume cosmological simulations is to perform zoom in simulations of selected halos (see Fig. 27). In zoom in simulations, typically one halo at a time is re-simulated at high resolution, this allows a stepping through of all the aspects of the physical models in order to understand what most affects the predictions, and improve/relax some aspect of the sub-grid models and/or add more ab-initio physics. Importantly, these type of simulations can be used to start testing different models for black hole seeds (based on proposed physical models) and improving some of the more approximate aspects of the BH treatment in large volume simulations. For example, our groups and others will be working on the treatment for the BH orbits, instead of glueing them to the potential minimum (which our simulations do currently). These kind of runs are used to inform large volume simulations of the effects of different physical implementations and ways they affect the merger events.

The resolution of BlueTides is sufficient to capture the large-scale processes of gas inflow which are dominant in the growth of the most supermassive black holes. The most important question in this work is to determine in detail how the gas actually enters the central regions of the galaxy, and what the role is of physics on smaller scales, such as *angular momentum/discs* (which we will be able to measure in the pc regions), gas drag and gas clumping. Zoomed resimulations are a way to improve the force resolution of the simulation so that it goes substantially below the current hundreds of parsec values (as in Bluetides), and also makes the mass resolution much finer. We started a program to re-simulate the most luminous quasars in the MB simulation [101]. In zoomed resimulations we can achieve $\sim$ parsec scales in spatial resolution and hence getting close to the gravitational radii of massive black holes.

One advantage also of these the zoomed re-simulations is that they can be run using different code, either SPH codes such as MP-Gadget and GASOLINE, or Adaptive Mesh Refinement Eulerian codes such as *RAMSES*, all starting but using the same initial conditions (Fig. 27). Using the two different hydrodynamic methods to tackle the formation of the same luminous quasars should offer convincing evidence of the gas inflow process at parsec scales.

By following the evolution of gas from the large scales down to the small scales (with appropriate zoom simulations; e.g., Feng et al. [101]) we find that in the early Universe the large black holes are fed by cold gas from large distances that is virtually unstoppable (even in the presence of AGN/supernova feedback) as it flows all the way into the black hole.

Other important zoomed runs, carried out by different teams have included and tested different models of BH feedback (jets as well as isotropic winds), and included radiative feedback based on direct hydrodynamic and radiation transfer in small scale accretion simulation studies.

6 Black Holes Grow: Modern Galaxies and Their Black Holes

Hopefully, from what we have described above it is clear that from a simulation/theoretical perspective we have now reached a major breakthrough. On the largest, modern high-performance computers it is now feasible to run computations with unprecedented volume and resolution. We can now aim at and start to formulate and develop a picture of the underlying physics regulating black hole growth/quasars and their role for understanding galaxy formation and feedback as well as their use in cosmology. Our ultimate goal would be to finally place firm constraints on the growth of black holes and their role in galaxy formation, understand the relation between BH activity and dark matter halo mass and promote their role as probes for cosmology. After 30 years of studying the quasar luminosity function and their clustering, we have now reached the juncture—both from the data side and the theory/simulation side—where an understanding of physics over large volumes at high resolution can truly advance our understanding of quasars and their roles in galaxy formation and cosmology.

6.1 Black Hole Mass—Galaxy Properties Relations: The Connection

In the local universe, the discovery of close relationships between the masses of supermassive black holes and several properties of their bulges such as the stellar mass stellar velocity dispersion ($M_{\rm BH} - \sigma$ relation, and other parameters have revolutionized our view of BHs, linking their growth to that of its host galaxy.

To understand the evolution of these relations at higher redshifts (mostly up to $z \sim 2-4$) observational studies rely on galaxies with active galactic nuclei for which BH mass estimates use the virial method. Some of these studies have found an evolution in which BH growth precedes galaxies but others favor no evolution, while others imply little or no evolution. The systematic uncertainties in the higher z measurements are still large and come from both the method for BH mass estimation and from measuring host galaxy properties.

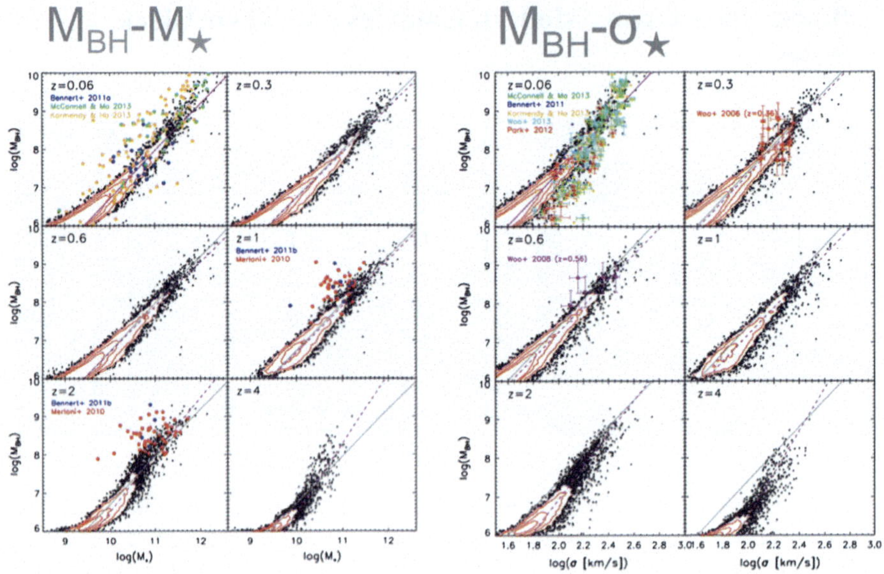

Fig. 28 An example of the results from simulations for the black hole—galaxy scaling relations [102]

A popular way to interpret these relationships is by assuming that supermassive BHs regulate their own growth and that of their hosts by coupling some (small) fraction of their energy output to their surrounding gas. AGN feedback heats and unbinds significant fractions of the gas and inhibits star formation. The scaling relations of black hole mass with the stellar properties of the host galaxies form a way to understand the importance and the effects of AGN feedback.

All the state-of-the-art cosmological hydrodynamical simulations of structure formation (*MassiveBlack-II* [51], Illustris [52], IllustrisTNG [75], EAGLE [49], etc.) have been used to investigate predictions for the galaxy-black hole relations $M_{\rm BH} - \sigma$, $M_{\rm BH} - M_{*,tot}$ and $M_{\rm BH} - L_{V,tot}$, relevant for the whole population of black holes and compare them to the observational constraints at $z = 0 - 2$. They are also many predictions from the simulations for the redshift evolution of these relations. All of the simulations that have been used to study the relations model AGN feedback and all obtain results that are consistent with the observed relations, and consistent with each other. They also imply very little evolution for any of these relations. The most interesting question that remains unanswered is with regards to the origin of the scatter in the relations. We show in Fig. 28 examples for these relations derived from a sub-set of recent simulations. Interestingly, simulations now have enough sophistication that we can mock up observations to the point that we can test the simulated population for possible selection biases in all types of observed relations [102, 103]. Indeed, we typically find that for samples selected on the basis of $M_{\rm BH}$ or M_* (and of size similar to those observed) the slopes can be steeper than for randomly selected samples. Such sample selection also biases toward finding stronger evolution with redshift than for a random sample as they tend to pick objects at the high-end of the relation (consistent with stronger evolution). This is relevant for some of the claims of evolution in the measurements. It is an interesting direction that simulations can indeed be used to pin down possible observational biases from actual physical evolution.

6.2 The QSO Luminosity Functions (QLF)

Even though the QLF has been studied for over 30 years we still do not understand the fundamental physical parameters that regulate its shape and evolution. Ultimately the evolution of the quasar luminosity function (QLF) is one of the basic cosmological measures providing insight into structure formation and its relation to black hole growth. The QLF is typically described by two power-law components: flatter and steeper at the faint and bright end respectively and with a break luminosity that evolves with redshift (luminosity density evolution). The bright-end slope also appears to evolve becoming flatter at the highest redshifts ($z = 5 - 6$), although the most recent measurements are hinting that this may not be the case [46, 104]. Theoretical investigation of the QLF has been done using semi-analytical models or halo models Since, by construction, these models do not self-consistently follow black hole growth, the quasar lightcurves (and luminosities) have to be calculated via imposed prescriptions and a number of parameters are introduced for quasar triggering, quasar lifetimes, etc. So while these models offers more flexibility for testing a variety of reasonable prescriptions and have produced promising results it is still ideal to complement these approaches with detailed hydrodynamic simulations. For example, in Fig. 29 we show a particular a view of two regions that contain two massive BHs but with very different accretion histories and luminosities, in the $z = 4$ output from one of our simulations (MBII). In particular, these frames show the underlying distribution of gas (color coded by temperature, red is hot and blue cold) with stars (in white) and black holes indicated by the diffraction spikes whose size is scaled by the QSO luminosity. Many things are evident from this picture. For example there is a clear distribution of quasar luminosities related to large scale properties, and effects of BH feedback (clearly seen as hot gas around the BHs). The same mass halos, can be found in different large scale environments hosting dramatically different AGNs (see also Fig. 29). We see clearly that the effects modeled in the simulations also directly provide for each BH a detailed prediction of its full lightcurve across cosmic history with high time resolution. This is directly predicted by the interaction of the cosmological gas supply and resulting BH feedback. An example of a lightcurve and the predictions of the associated luminosity functions derived from such a population is shown in Fig. 30 (in black).

If we relate this to the QLFs, while the bright end QLF may inform us about feedback, the faint-end where BH are already mostly self-regulated should inform us about gas supply. In our simulations we typically find that although the low (high) luminosity ranges of the faint-end QLF are dominated by low (high) mass black holes, a wide range of black hole masses still contributes to any given luminosity

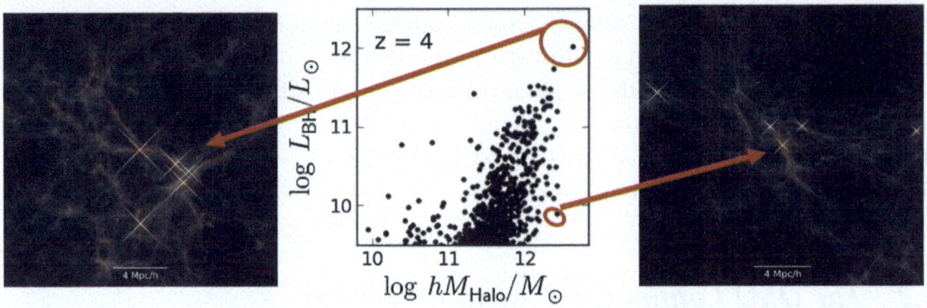

Fig. 29 The effect of environment on SMBH mass. Two galaxies from the MassiveBlack II simulation are shown, along with their location in the scatter plot of BH mass versus halo mass. Both halos have a mass $\sim 10^{12} M_{\odot}$, but have experienced very different gas inflow and hence black hole growth [101]

range. The faint-end of the QLF can indeed be formed by quasars radiating well below their peak luminosities, rather than by quasars with low peak luminosities. This is consistent with the complex lightcurves of black holes, which show that any given black hole can undergo significant changes in its luminosity and hence (while its mass always grows) it can occupy different parts of the LF. The complex light curve, and the resulting effects on the LFs are a result of the detailed hydrodynamics followed in the simulations.

6.3 *Quasar Clustering*

Clustering measurements provide the means to better understand the relation between quasars, their hosts and the underlying dark matter distribution, as well as to allow estimates of quasar lifetimes once coupled with QLF constraints, e.g. [105, 106]. With clustering we learn about the co-evolution of quasars, mergers and red galaxies (and hence the efficiency of feedback). For example, strong clustering would suggest that quasars should reside in massive halos. If so, they should be rare and in order to reproduce the quasar luminosity density, they must have long lifetimes. Conversely, low spatial correlations would suggest more common quasars, and thus shorter quasar lifetimes (see Fig. 30).

Our MassiveBlack simulation has been used to compare to constraints on clustering [103] showing that quasar hosts at high-z are consistent with the level of SDSS quasar clustering. This is an important starting point which gives us some confidence to further pursue lower redshifts—towards the peak epoch of quasar activity and overlap with the redshift range of BOSS and eBOSS quasars. Because of the detailed information and lightcurves and QLF we have from the simulations (e.g. Fig. 30) we can directly derive host halo masses for a given luminosity range and translate that into predicted quasar clustering as a function of z or luminosity.

Examples—extrapolated below $z = 4.75$ using quasar halo occupation distribution (HOD) formalism—are shown in Fig. 31. It is evident that by taking into account the evolution of quasar lightcurves, the quasar lifetime, t_Q, is strongly dependent on the luminosity considered (see for example two estimates of t_Q in Fig. 30). This implies much longer quasar lifetimes at low luminosity versus high, and, in turn, a particular trend with redshift. Models built on a simple lightcurve would end up ascribing the luminosity dependence of clustering simply to different halo mass hosts—thereby suggesting typically a much stronger evolution. By matching to appropriate luminosity cuts from the ongoing BOSS and upcoming eBOSS analyses and pushing large volume simulations all the way to $z = 2$, we will be able to directly compare with the data. We will be able to discriminate between different models, and directly constrain duty cycles and the effects of gas inflows and feedback in regulating quasar active phases.

6.4 AGN Feedback and Cosmology

AGN feedback is key for reproducing the global evolution of the stellar mass function, galaxy luminosity functions. With the simulations we are also able to firmly predict the quasar bias at all scales. The newest surveys coupled with these predictions will allow us to determine how quasars probe large scale structure from very small scales to Cosmic Microwave background scales. This work will be important for planning future surveys. Current and forthcoming astronomical surveys (SDSS, LSST etc.) rely on observed galaxy properties to constrain the effects of dark matter and dark energy with increasing accuracy. The fundamental challenge is that galaxy formation involves a complicated blend of different physics that is non-linearly coupled on a

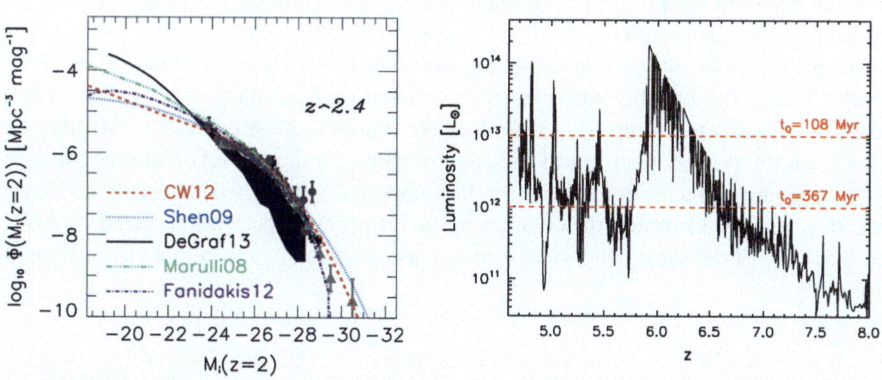

Fig. 30 Left: predictions for the quasar (bolometric) luminosity functions from MassiveBlack in black and the SDSS/BOSS QLFs. Right: An example of a bolometric lightcurve from one quasar (black hole) from the MassiveBlack simulation. The dashed blue line shows the Eddington luminosity and the red quasar lifetimes at corresponding luminosities

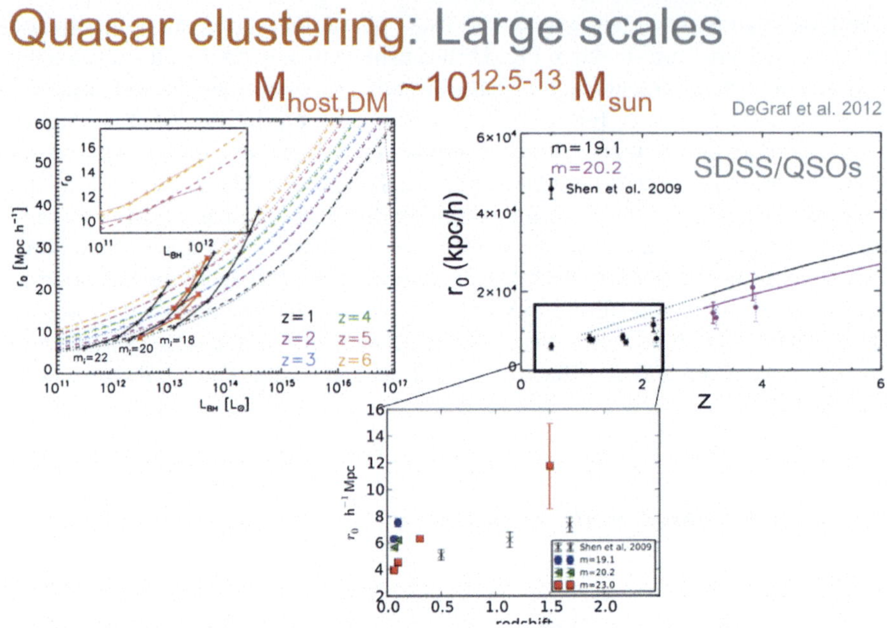

Fig. 31 Dependence of quasar correlation length, r_0 on luminosity from the simulations (fit with HOD quasar model) compared with SDSS measurements

wide range of scales, leading to extremely complex dynamics. The required percent level accuracy for LSST can only be reached through a much better understanding of galaxy formation in direct cosmological simulations. BH are important and perhaps the most important piece of baryonic physics we need to understand for constraining dark energy with upcoming LSST Weak lensing shear and this depends heavily on calibration from simulations.

As an example, using cosmological simulations it has been shown that AGN feedback may modify the matter power spectrum on fairly large scales [108, 109], reaching corrections of up to ~30% at wave numbers as small as $k \sim 10$ Mpc^{-1}. Effects on the power spectrum of this size must be accurately taken into account in order to achieve precise constraints on dark energy through missions such as Euclid or LSST. As Fig. 32 shows, the changes in the DM power spectrum induced by AGN feedback are of the same order of magnitude as those due to a different cosmological mode [75].

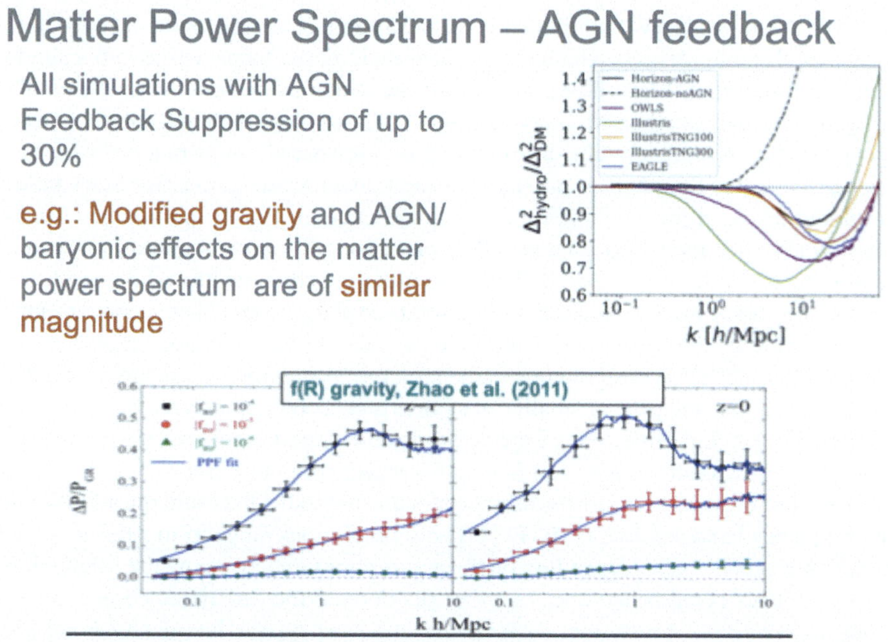

Fig. 32 Top right panel shows the ratio of the matter power spectrum in hydrodynamic simulations which include BH/feedback modelling versus the pure dark matter counterparts adapted from [75]. In the bottom panel the ratio of matter power spectrum for modified gravity models [107]

7 Massive Black Holes and Galaxy Mergers

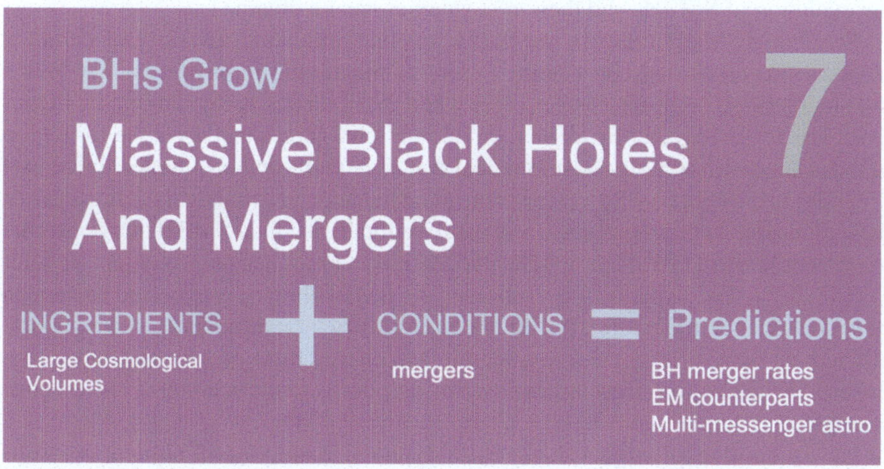

While black holes grow predominantly via accretion, a second mode of black hole growth is through mergers which occur when dark matter halos merge into a single halo, such that their black holes fall toward the center of the new halo, eventually merging with one another. Mergers of massive black holes are then a natural consequence of our current hierarchical structure formation paradigm. In much of what has been discussed in this chapter and in our current understanding, massive black holes form and reside at the centers of galaxies and hence they grow and merge closely intertwined with their host galaxies. The presence of luminous quasars observed within the first billion years of the Universe highlights that the black hole seeds for the massive black hole population were assembled at the cosmic dawn, concurrently with the time of the formation of the first galaxies. In our standard Lambda Cold Dark Matter (ΛCDM) cosmology cosmic structure formation occurs hierarchically by the continuous merging of smaller structures and accretion of surrounding matter. SMBHs growth and evolution is expected to follow a similar process in which black hole seeds grow both though accretion and mergers with other BHs.

In the last decade major efforts have been made to predict the event rate of GWs in the frequency band of LISA [110, 111]. These predictions range from a few to a few hundred events per year, depending on the assumptions underpinning the calculation of the SMBHs coalescence rate. Early works derived the SMBH coalescence rate from observational constraints such as the observed quasar luminosity function, whilst more recent studies have utilised semi-analytical galaxy formation models and/or hybrid models that combine cosmological N-body simulations with semi-analytical recipes for the SMBH dynamics [112–117].

In contrast to semi-analytic models, hydrodynamical simulations follow the dynamics of the cosmic gas by direct numerical integration of the equations of hydrodynamics, capturing non-linear processes that cannot be described by simple mathematical approximations. Hence a more complete and consistent picture of the evolution of SMBHs and their host galaxies can be obtained.

Predicting SMBH mergers inevitably involves following a variety of complex physical processes that cover many orders of magnitude in physical scale. Black hole mergers occur at sub-parsec scales when two galaxies within large dark matter halos are driven together by large scale gravitational forces that drive the formation of the cosmic web at >Mpc cosmological scales. After the galaxy merger, the central SMBHs are brought near the center of the main halo due to dynamical friction against the dark matter, background stars, and gas. Eventually the final SMBH merger occurs via the emission of GWs. For reviews of SMBH dynamics in galaxy mergers we refer to [118–120]. The dynamical evolution of the SMBH binary is expected to happen fast (coalescence timescale 10–100 Myrs) in gas rich environments, thanks to the efficient dissipation of angular momentum and energy from the binary. Conversely, three-body interactions slow things down in gas poor systems (leading to coalescence timescales ~Gyrs).

Newly developed, large volume hydrodynamic cosmological simulations self-consistently combine the processes of structure formation at cosmological scales with the physics of smaller, galaxy scales and thus capture our most realistic understanding of black holes and their connection to galaxy formation they thus provide the most accurate prediction for MBHB merger rates.

LISA will detect mergers of black holes in the mass range of $10^4\,M_\odot - 10^7\,M_\odot$ out to $z = 20$ and measure their mass, spin and distance.

The launch of LISA will extend the GW window to low frequencies, opening new investigations into dynamical processes involving these massive black hole binaries (MBHB). MBHB are also the primary multimessenger astrophysics sources. The GW events will be accompanied by electromagnetic (EM) counterparts and, since information carried electromagnetically is complementary to that carried gravitationally, a great deal can be learnt about an event and its environment (binary AGN and its host galaxy and beyond) across cosmic history as it becomes possible to measure both forms of radiation in concert.

Devising observing strategies for the new multimessenger astrophysics LISA opens up will significantly benefit from predictions of EM counterparts of binary AGN and SMBH host galaxies. It will require 'full-physics' hydrodynamical cosmological simulations with sufficient resolution and volume. One of the major goal objectives of LISA is to trace the origin, growth and merger history of massive black holes across cosmic ages.

The state-of-the art multi-scale hydrodynamical simulations that we have been describing include different implementations for BH growth and associated feedback. They can be used used to predict SMBH mergers rates. and perform accurate studies of the predictions for SMBH merger rates using *MassiveBlackII (MBII)*, *Illustris* and *IllustrisTNG* simulations to clarify any discrepancies and where models converge or need improvements. Excitingly they can also provide corresponding EM counterparts and host galaxies for the MBHBs that can be observed in future and upcoming facilities. Host galaxy identification of MBHB provides unique information on galaxy-BH coevolution (and precise determination of the distance-redshift relation). The first LISA detections of massive black hole mergers will mobilize global astronomical resources and be an astronomical event of enormous excitement. The mock catalogs and synthetic observations that one can obtain should be

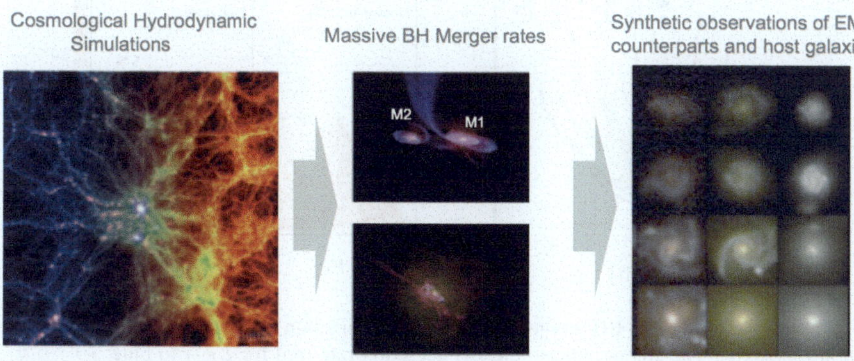

Fig. 33 From state-of-the-art cosmological simulations to massive BH merger events, to mission specific synthetic EM counterpart observations

able bring traditional astronomers into the LISA community and begin LISA science with MBHM even before LISA is launched (Fig. 33).

In contrast to semi-analytic models, hydrodynamical simulations follow the dynamics of the cosmic gas by direct numerical integration of the equations of hydrodynamics, capturing non-linear processes that cannot be described by simple mathematical approximations. Hence a more complete and consistent picture of the evolution of SMBHs and their host galaxies can be obtained.

To illustrate some of the data products that we have available in the simulations we show some preliminary analyses. For each SMBH merger that takes place in the simulations we store the mass of both SMBHs, $M1$ and $M2$, and the redshift z at which the merger event takes place. Figure 34 we show the 2D histogram of the mass of each BH member for all the mergers in both the Illustris and MBII simulations considered here. The total number of BH mergers in each simulation model is indicated in the figure.

After compiling a database of massive binary BH candidates from the large cosmological simulations, we can construct and organize predictions for their host galaxy morphologies and AGN signatures. The goal is to enable multi-messenger astronomy with LISA sources via detailed comparisons between putative LISA events and telescope data that would illuminate properties of the EM counterparts and histories of their host galaxies.

Following the dynamics of galaxy mergers with their black holes we can to estimate the incidence of *dual AGN* (at least down to typical separation of a few kpc, at which we still can resolve dynamical friction directly) and the detectability of these binary systems. Correspondingly we have detailed properties of the stellar distribution of the host galaxies, with age, metallicities, star formation rates and associated morphologies. As illustrated in Fig. 35 we will be able to statistically *characterize the type of galaxy for given minor/major BH merger events across redshift*. Even with LISA more than a decade away, our aim would be to predict which future facilities (JWST, Luvoir, OST) will be needed to to observe the type of galaxy which will be

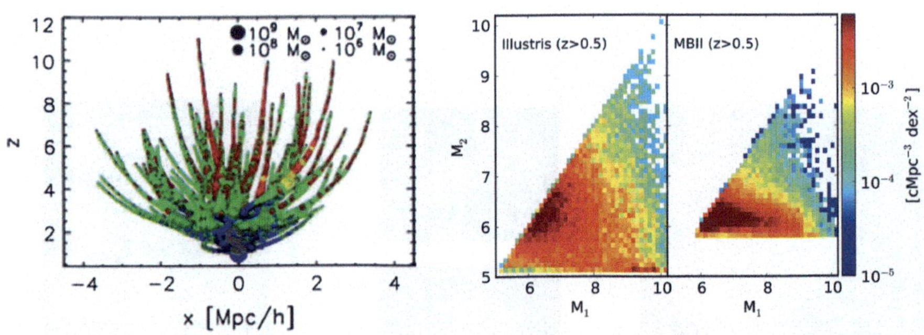

Fig. 34 Left: An illustrative BH merger history tree as a function of redshift z on the y-axis and position (one axis). The colors indicate the BH accretion rate (red for L_{Edd} to blue: $10^{-4}L_{edd}$. Right: distribution of BH masses ($M1$ is the primary and $M2$ the secondary involved in mergers. Illustris and MBH2 are compared

MBHB hosts out to high redshifts. For example, the $z = 5$ example shown in Fig. 35 is the highest redshift host galaxy of a MBHB event in Illustris that will be detectable by JWST (as at these redshifts a MBHB typically involves lower masses) [122]. We will use make predictions for what Luvoir and OST should see as this will be particularly relevant for probing the BH occupation fraction up to high redshifts. The process utilizes public software such as Sunrise [123] and YT [124]. These tools have been used to develop a Synthetic Deep Fields High-Level Science Product hosted by the Mikulski Archive for Space Telescopes (MAST): https://archive.stsci.edu/prepds/illustris/ (doi: 10.17909/T98385).

In Fig. 35, we show mock JWST images and mass assembly histories of several galaxies hosting MBHB sources in Illustris. We selected these sources to span a range of redshifts, masses, and mass ratios (note the detailed BH merger tree for $z = 3$ and $z = 2$ host galaxies are also shown in Fig. 34. Such products can be used to characterize galaxy morphology and AGN activity which may indicate recent (bulges) or ongoing (companions, tails, etc) merging activity and therefore link (in simulations) the population of LISA sources to the story of how galaxy populations assembled. Figure 36 shows the time evolution of a single such source over ~500 Myr. In this evolution, a minor galaxy merger delivers a MBH with $M2/M1 \sim 100$ to the primary host ($M1 \sim 10^9$), and these BHs merge near the time shown in the 3rd panel.

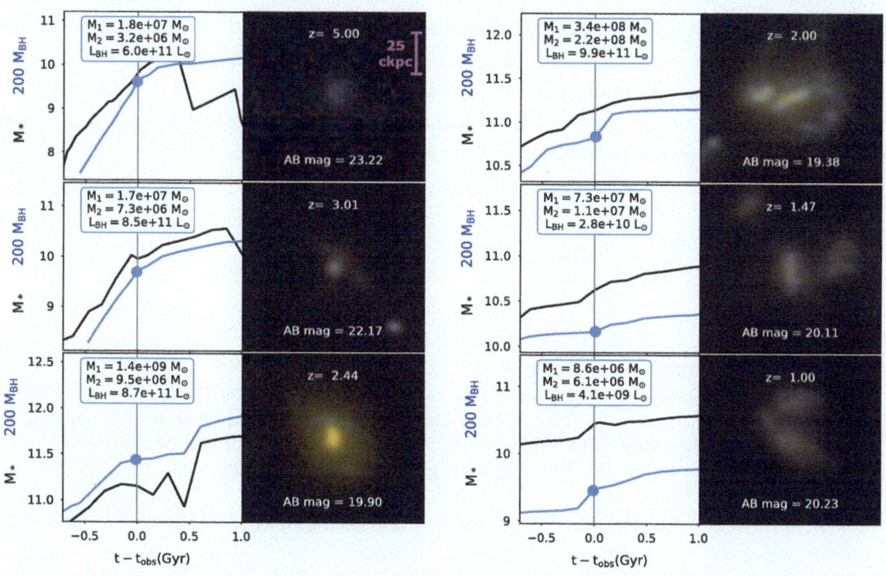

Fig. 35 Six example potential LISA sources from a catalog of MBHB mergers in Illustris, spanning mass, mass ratio, and redshift. These plots represent a small window on the M_* and $M_{\rm BH}$ history of the main progenitor host in the left panel. The right panel shows simulated JWST-Nircam images of the hosts [121]. Many of these sources are associated with obvious galaxy mergers, and there is substantial diversity in the host morphology, luminosity, and color

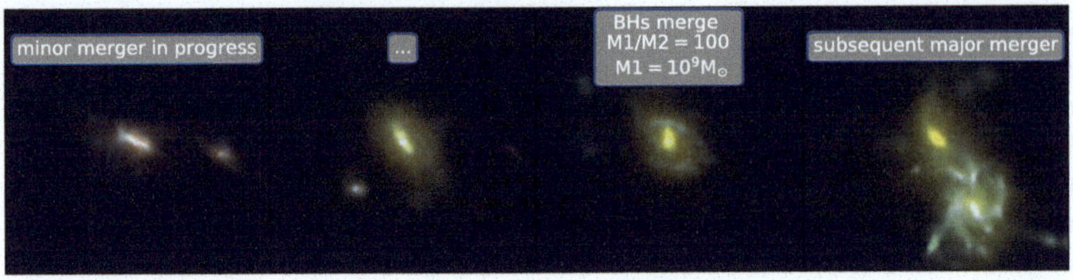

Fig. 36 Roughly $500Myr$ of evolution of the host galaxy for the $z = 2.44$ source in Fig. 35 (3rd panel). Labels describe the visible assembly processes acting on the galaxy and MBHB. The primary MBH has AGN emission visible as a bright blue point source in the first 3 panels. Each panel has a fixed 50 physical kpc field of view [122]

We can see that this MBHB merger occurs during a period of rapid galaxy assembly in the host.

Following the dynamics of galaxy mergers with their black holes we should be able to estimate the incidence of *dual AGN* (at least down to typical separation of a few kpc, at which we still can resolve dynamical friction directly) and the detectability of these binary systems [125].

Currently a few teams have been able to carry out impressive MHD simulations of circumbinary disks [126–128] around relativistic binary BHs which are now starting to produce detailed EM counterpart signatures for these events. In the near future it will be possible to use large scale simulations to provide reasonable 'initial conditions' of the gas environments for these smaller scales around the BHs at the time of mergers that the detailed simulations could use to derive realistic *EM signatures for a given mass ratio event in a given environment/galaxy host.*

References

1. Carr, B., Kühnel, F., Sandstad, M.: Primordial black holes as dark matter. Phys. Rev. D 94, 083504 (2016). https://link.aps.org/doi/10.1103/PhysRevD.94.083504
2. Chapline, G.F.: Cosmological effects of primordial black holes. Nature **253**, 251–252 (1975)
3. Garcia-Bellido, J., Linde, A., Wands, D.: Density perturbations and black hole formation in hybrid ination. Phys. Rev. D **54**, 6040–6058 (1996). arXiv:astro-ph/9605094 [astro-ph]
4. Garcia-Bellido, J., Clesse, S.: Black holes from the beginning of time. Sci. Am. **317**, 38–43 (2017)
5. Carr, B.J., Rees, M.J.: How large were the first pregalactic objects? MNRAS **206**, 315–325 (1984)
6. Bird, S., et al.: Did LIGO detect dark matter? Phys. Rev. Lett. **116**, 201301 (2006). arXiv:1603.00464 [astro-ph.CO]
7. Hawking, S.W.: Particle creation by black holes. Commun. Math. Phys. **43**, 199–220 (1975)
8. Crawford, M., Schramm, D.N.: Spontaneous generation of density perturbations in the early Universe. Nature **298**, 538–540 (1982)
9. Hawking, S.W.: Black holes from cosmic strings. Phys. Lett. B **231**, 237–239 (1989)
10. Polnarev, A., Zembowicz, R.: Formation of primordial black holes by cosmic strings. Phys. Rev. D **43**, 1106–1109 (1991)

11. Clesse, S., Garcia-Bellido, J.: Massive primordial black holes from hybrid ination as dark matter and the seeds of galaxies. Phys. Rev. D **92**, 023524 (2015). arXiv:1501.07565 [astro-ph.CO]
12. Cirelli, M.: Dark matter indirect searches: charged cosmic rays. J. Phys. Conf. Ser. **718**, 022005 (2016)
13. Cirelli, M., Taoso, M.: Updated galactic radio constraints on dark matter. J. Cosmol. Astro-Part. Phys. **041**, (2016). arXiv:1604.06267 [hep-ph]
14. Ali-Haimoud, Y., Kovetz, E.D., Kamionkowski, M.: Merger rate of primordial black-hole binaries. Phys. Rev. D **96**, 123523 (2017). arXiv:1709.06576 [astro-ph.CO]
15. Kovetz, E.D.: Probing primordial black hole dark matter with gravitational waves. Phys. Rev. Lett. **119**, 131301 (2017). arXiv:1705.09182 [astro-ph.CO]
16. Regan, J.A., Haehnelt, M.G.: Pathways to massive black holes and compact star clusters in pre-galactic dark matter haloes with virial temperatures >10000K. MNRAS **396**, 343–353 (2009). arXiv:0810.2802
17. Abel, T., Bryan, G.L., Norman, M.L.: The formation of the first star in the universe. Science **295**, 93–98 (2002)
18. Johnson, J.L., Bromm, V.: The aftermath of the first stars: massive black holes. MNRAS **374**, 1557–1568 (2007)
19. Madau, P., Rees, M.J.: Massive black holes as population III remnants. ApJ **551**, L27–L30 (2001). arXiv:astro-ph/0101223
20. Heger, A., Woosley, S.E., Fryer, C.L., Langer, N.: Massive Star Evolution Through the Ages in From Twilight to Highlight: The Physics of Supernovae, Hillebrandt, W., Leibundgut, B. (eds.), vol. 3 (2003). arXiv:astro-ph/0211062
21. Begelman, M.C., Volonteri, M., Rees, M.J.: Formation of supermassive black holes by direct collapse in pre-galactic haloes. MNRAS **370**, 289–298 (2006). arXiv:astro-ph/0602363
22. Ferrara, A., Salvadori, S., Yue, B., Schleicher, D.: Initial mass function of intermediate-mass black hole seeds. MNRAS **443**, 2410–2425 (2014). arXiv:1406.6685
23. Latif, M.A., Schleicher, D.R.G., Schmidt, W., Niemeyer, J.: Black hole formation in the early Universe. MNRAS **433**, 1607–1618 (2013). arXiv: 1304.0962 [astro-ph.CO]
24. Lodato, G., Natarajan, P.: Supermassive black hole formation during the assembly of pre-galactic discs. MNRAS **371**, 1813–1823 (2006). arXiv:astroph/0606159
25. Rees, M.J.: Accretion and the quasar phenomenon. Phys. Sci. **17**, 193–200 (1978)
26. Clark, P.C., et al.: The formation and fragmentation of disks around primordial protostars. Science **331**, 1040 (2011). arXiv:1101.5284 [astro-ph.CO]
27. Schneider, R., Omukai, K., Inoue, A.K., Ferrara, A.: Fragmentation of star-forming clouds enriched with the first dust. MNRAS **369**, 1437–1444 (2006). arXiv:astro-ph/0603766
28. Begelman, M.C., Rees, M.J.: The fate of dense stellar systems. MNRAS **185**, 847–860 (1978)
29. Devecchi, B., Volonteri, M.: Formation of the first nuclear clusters and massive black holes at high redshift. ApJ **694**, 302–313 (2009). arXiv:0810.1057
30. Yajima, H., Khochfar, S.: The role of stellar relaxation in the formation and evolution of the first massive black holes. MNRAS **457**, 2423–2432 (2016). arXiv:1507.06701
31. Volonteri, M., Bellovary, J.: Black holes in the early Universe. Rep. Prog. Phys. **75**, 124901 (2012). arXiv:1209.2243
32. Turk, M.J., Abel, T., O'Shea, B.: The formation of population III binaries from cosmological initial conditions. Science **325**, 601 (2009). arXiv:0907.2919 [astro-ph.CO]
33. Trenti, M., Stiavelli, M., Michael Shull, J.: Metal-free gas supply at the edge of reionization: late-epoch population III star formation. ApJ **700**, 1672–1679 (2009). arXiv:0905.4504 [astro-ph.CO]
34. Greif, T.H., et al.: Formation and evolution of primordial protostellar systems. MNRAS **424**, 399–415 (2012)
35. Hirano, S., et al.: One hundred first stars: protostellar evolution and the final masses. ApJ **781**, 60 (2014). arXiv:1308.4456 [astro-ph.CO]
36. Latif, M.A., Schleicher, D.R.G., Spaans, M.: The implications of dust for high-redshift protogalaxies and the formation of binary disks. A & A **540**, A101 (2012). arXiv:1110.4256 [astro-ph.CO]

37. Regan, J.A., et al.: Rapid formation of massive black holes in close proximity to embryonic protogalaxies. Nat. Astron. **1**, 0075 (2017). arXiv:1703.03805 [astro-ph.GA]
38. Wise, J.H., et al. Formation of massive black holes in rapidly growing pregalactic gas clouds. Nature **566**, 85–88 (2019). arXiv:1901.07563
39. Hirano, S., Hosokawa, T., Yoshida, N., Kuiper, R.: Supersonic gas Streams enhance the formation of massive black holes in the early universe. Science **357**, 1375–1378 (2017). arXiv:1709.09863 [astro-ph.CO]
40. Chon, S., Hosokawa, T., Yoshida, N.: Radiation hydrodynamics simulations of the formation of direct-collapse supermassive stellar systems. MNRAS **475**, 4104–4121 (2018). arXiv:1711.05262 [astro-ph.GA]
41. Mayer, L., Kazantzidis, S., Escala, A., Callegari, S.: Direct formation of supermassive black holes via multi-scale gas inows in galaxy mergers. Nature **466**, 1082–1084 (2010)
42. Mayer, L., Bonoli, S.: The route to massive black hole formation via merger-driven direct collapse: a review. Rep. Progr. Phys. **82**, 29 (2019). arXiv:1803.06391
43. Mayer, L., et al.: Direct formation of supermassive black holes in metalenriched gas at the heart of high-redshift galaxy mergers. ApJ **466**, 51–65 (2015). arXiv:1411.5683
44. Woods, T., et al.: Titans of the Early Universe: The Prato Statement on the Origin of the First Supermassive Black Holes. Publications of the Astronomical Society of Australia, vol. 38 (2019). arXiv:eprintarXiv:1810.12310
45. Ba nados, E., et al.: An 800-million-solar-mass black hole in a significantly neutral Universe at a redshift of 7.5. Nature **553**, 473–476 (2018). arXiv:1712.01860
46. Fan, X., et al.: The first luminous quasars and their host galaxies (2019). arXiv e-prints. arXiv:1903.04078
47. Fan, X., et al.: A survey of $z > 5.7$ quasars in the sloan digital sky survey. IV. Discovery of seven additional quasars. AJ **131**, 1203–1209 (2006). arXiv:astro-ph/0512080
48. Spergel, D.N., et al.: Three-year Wilkinson microwave anisotropy probe (WMAP) observations: implications for cosmology. ApJS **170**, 377–408 (2007). arXiv:astro-ph/0603449
49. Crain, R.A., et al.: The EAGLE simulations of galaxy formation: calibration of subgrid physics and model variations. MNRAS **450**, 1937–1961 (2015). arXiv:1501.01311
50. Di Matteo, T., Croft, R.A.C., Feng, Y., Waters, D., Wilkins, S.: The origin of the most massive black holes at high-z: BlueTides and the next quasar frontier. MNRAS **467**, 4243–4251 (2017). arXiv:1606.08871
51. Khandai, N., et al.: The Massive black-II simulation: the evolution of haloes and galaxies to z 0. MNRAS **450**, 1349–1374 (2015). arXiv:1402.0888
52. Vogelsberger, M., et al.: Introducing the Illustris Project: simulating the coevolution of dark and visible matter in the Universe. MNRAS **444**, 1518–1547 (2014). arXiv:1405.2921
53. Di Matteo, T., et al.: Cold flows and the first quasars. ApJ **745**, L29 (2012). arXiv:1107.1253 [astro-ph.CO]
54. Springel, V.: The cosmological simulation code GADGET-2. MNRAS **364**, 1105–1134 (2005)
55. Feng, Y., et al.: The formation of milky way-mass disk galaxies in the first 500 million years of a cold dark matter universe. ApJ **808**, L17 (2015). arXiv:1504.06618
56. Barnes, J., Hut, P.: A hierarchical O(N log N) force-calculation algorithm. Nature **324**, 446–449 (1986)
57. Hopkins, P.F., et al.: The evolution in the faint-end slope of the quasar luminosity function. ApJ **639**, 700–709 (2006). arXiv:astro-ph/0508299
58. Springel, V.: E pur si muove: Galilean-invariant cosmological hydrodynamical simulations on a moving mesh. MNRAS **401**, 791–851 (2010). arXiv:0901.4107 [astro-ph.CO]
59. Bauer, A., Springel, V.: Subsonic turbulence in smoothed particle hydrodynamics and moving-mesh simulations. MNRAS **3102**, (2012). arXiv:1109.4413 [astro-ph.CO]
60. Genel, S., et al.: Following the ow: tracer particles in astrophysical uid simulations. MNRAS **435**, 1426–1442 (2013). arXiv:1305.2195 [astro-ph.IM]
61. Nelson, D., et al.: Moving mesh cosmology: tracing cosmological gas accretion. MNRAS **429**, 3353–3370 (2013). arXiv:1301.6753 [astro-ph.CO]

62. Sijacki, D., Vogelsberger, M., Kereš, D., Springel, V., Hernquist, L.: Moving mesh cosmology: the hydrodynamics of galaxy formation. MNRAS **424**, 2999–3027 (2012). arXiv:1109.3468 [astro-ph.CO]
63. Torrey, P., Vogelsberger, M., Sijacki, D., Springel, V., Hernquist, L.: Moving-mesh cosmology: properties of gas discs. MNRAS **427**, 2224–2238 (2012). arXiv:1110.5635 [astro-ph.CO]
64. Vogelsberger, M., Sijacki, D., Kereš, D., Springel, V., Hernquist, L.: Moving mesh cosmology: numerical techniques and global statistics. MNRAS **425**, 3024–3057 (2012). arXiv:1109.1281 [astro-ph.CO]
65. Vogelsberger, M.: Cosmological simulations of dark matter in APS. Meeting Abstr. **R10**, 003 (2015)
66. Di Matteo, T., Springel, V., Hernquist, L.: Energy input from quasars regulates the growth and activity of black holes and their host galaxies. Nature **433**, 604–607 (2005)
67. Springel, V., et al.: Simulations of the formation, evolution and clustering of galaxies and quasars. Nature **435**, 629–636 (2005)
68. DeGraf, C., Sijacki, D.: Black hole clustering and duty cycles in the Illustris simulation. MNRAS **466**, 3331–3343 (2017). arXiv:1609.06727
69. Sijacki, D., et al.: The Illustris simulation: the evolving population of black holes across cosmic time. MNRAS **452**, 575–596 (2015). arXiv:1408.6842
70. Weinberger, R., et al.: Simulating galaxy formation with black hole driven thermal and kinetic feedback. MNRAS **465**, 3291–3308 (2017). arXiv:1607.03486
71. Genel, S., et al.: Introducing the Illustris project: the evolution of galaxy populations across cosmic time. MNRAS **445**, 175–200 (2014). arXiv:1405.3749
72. Marinacci, F., et al.: First results from the IllustrisTNG simulations: radio haloes and magnetic fields (2017). ArXiv e-prints. arXiv:1707.03396
73. Naiman, J.P., et al.: First results from the IllustrisTNG simulations: a tale of two elements - chemical evolution of magnesium and europium. MNRAS **477**, 1206–1224 (2018). arXiv:1707.03401
74. Pillepich, A., et al.: Simulating galaxy formation with the IllustrisTNG model. MNRAS **473**, 4077–4106 (2018). arXiv:1703.02970
75. Springel, V., et al.: First results from the IllustrisTNG simulations: matter and galaxy clustering. MNRAS **475**, 676–698 (2018). arXiv:1707.03397
76. Bhowmick, A.K., Di Matteo, T., Feng, Y., Lanusse, F.: The clustering of $z > 7$ galaxies: predictions from the BLUETIDES simulation. MNRAS **474**, 5393–5405 (2018). arXiv:1707.02312
77. Huang, K.-W., Di Matteo, T., Bhowmick, A.K., Feng, Y., Ma, C.-P.: BLUETIDES simulation: establishing black hole-galaxy relations at highredshift. MNRAS (2018). arXiv:1801.04951
78. Ni, Y., Di Matteo, T., Feng, Y., Croft, R.A.C., Tenneti, A.: Gas outows from the z = 7.54 quasar: predictions from the BLUETIDES simulation. MNRAS **481**, 4877–4884 (2018). arXiv:1806.00184
79. Tenneti, A., Di Matteo, T., Croft, R., Garcia, T., Feng, Y.: The descendants of the first quasars in the BlueTides simulation. MNRAS **474**, 597–603 (2018). arXiv:1708.03373
80. Wilkins, S.M., et al.: The properties of the first galaxies in the BlueTides simulation. MNRAS **469**, 2517–2530 (2017). arXiv:1704.00954
81. Feng, Y., et al.: The BlueTides simulation: first galaxies and eionization. MNRAS **455**, 2778–2791 (2016). arXiv:1504.06619
82. Gebhardt, K., et al.: A relationship between nuclear black hole mass and galaxy velocity dispersion. ApJ **539**, L13–L16 (2000)
83. Gültekin, K., et al.: The M-$_\sigma$ and M-L relations in galactic bulges, and determinations of their intrinsic scatter. ApJ **698**, 198–221 (2009). arXiv:0903.4897 [astro-ph.GA]
84. Häring, N., Rix, H.-W.: On the black hole mass-bulge mass relation. ApJ **604**, L89–L92 (2004). arXiv:astro-ph/0402376
85. Kormendy, J., Ho, L.C.: Coevolution (Or Not) of supermassive black holes and host galaxies. ARA & A **51**, 511–653 (2013). arXiv:1304.7762 [astro-ph.CO]
86. Magorrian, J., et al.: The demography of massive dark objects in galaxy centers. AJ **115**, 2285–2305 (1998)

87. McConnell, N.J., Ma, C.-P.: Revisiting the scaling relations of black hole masses and host galaxy properties. ApJ **764**, 184 (2013). arXiv:1211.2816
88. Reines, A.E., Volonteri, M.: Relations between central black hole mass and total galaxy stellar mass in the local universe. ApJ **813**, 82 (2015). arXiv:1508.06274
89. Tremaine, S., et al.: The slope of the black hole mass versus velocity dispersion correlation. ApJ **574**, 740–753 (2002)
90. Bower, R.G., et al.: Breaking the hierarchy of galaxy formation. MNRAS **370**, 645–655 (2006). arXiv:astro-ph/0511338
91. Ciotti, L., Ostriker, J.P., Proga, D.: Feedback from central black holes in elliptical galaxies. I. Models with either radiative or mechanical feedback but not both. ApJ **699**, 89–104 (2009). arXiv:0901.1089 [astro-ph.GA]
92. Croton, D.J., et al.: The many lives of active galactic nuclei: cooling ows, black holes and the luminosities and colours of galaxies. MNRAS **365**, 11–28 (2006)
93. Di Matteo, T., Colberg, J., Springel, V., Hernquist, L., Sijacki, D.: Direct cosmological simulations of the growth of black holes and galaxies. ApJ **676**, 33–53 (2008)
94. King, A.: Black Holes, Galaxy Formation, and the $M_{BH} - \sigma$ Relation. ApJ **596**, L27–L29 (2003)
95. Silk, J., Rees, M.J.: **331**, L1–L4 (1998)
96. Hirschmann, M., et al.: On the evolution of the intrinsic scatter in black hole versus galaxy mass relations. MNRAS **407**, 1016–1032 (2010). arXiv:1005.2100 [astro-ph.GA]
97. Jahnke, K., Macció, A.V.: The non-causal origin of the black- holegalaxy scaling relations. ApJ **734**, 92 (2011). arXiv:1006.0482 [astro-ph.CO]
98. Oesch, P.A., et al.: A remarkably luminous galaxy at z = 11.1 measured with hubble space telescope grism spectroscopy. ApJ **819**, 129 (2016). arXiv:1603.00461
99. Waters, D., Di Matteo, T., Feng, Y., Wilkins, S. M., Croft, R.A.C.: Forecasts for the WFIRST high latitude survey using the BlueTides simulation. MNRAS **463**, 3520–3530 (2016). arXiv:1605.05670
100. Waters, D., et al.: Monsters in the dark: predictions for luminous galaxies in the early Universe from the BLUETIDES simulation. MNRAS **461**, L51–L55 (2016). arXiv:1604.00413
101. Feng, Y., Di Matteo, T., Croft, R., Khandai, N.: High-redshift supermassive black holes: accretion through cold ows. MNRAS **440**, 1865–1879 (2014). arXiv:1312.1391
102. DeGraf, C., et al.: Scaling relations between black holes and their host galaxies: comparing theoretical and observational measurements, and the impact of selection effects. MNRAS **454**, 913–932 (2015). arXiv:1412.4133
103. DeGraf, C., et al.: Early black holes in cosmological simulations: luminosity functions and clustering behaviour. MNRAS **424**, 1892–1898 (2012). arXiv:1107.1254 [astro-ph.CO]
104. Giallongo, E., et al.: Faint AGNs at $z > 4$ in the CANDELS GOODS-S field: looking for contributors to the reionization of the Universe. A&A **578**, A83 (2015). arXiv:1502.02562
105. Haiman, Z., Hui, L.: Constraining the lifetime of quasars from their spatial clustering. ApJ **547**, 27–38 (2001). arXiv:astro-ph/0002190
106. Martini, P., Weinberg, D.H.: Quasar clustering and the lifetime of quasars. ApJ **547**, 12–26 (2001). arXiv:astro-ph/0002384
107. Zhao, G.-B., Li, B., Koyama, K.: N-body simulations for f(R) gravity using a self-adaptive particle-mesh code. Phys. Rev. D **83**, 044007 (2011). arXiv:1011.1257 [astro-ph.CO]
108. Semboloni, E., Hoekstra, H., Schaye, J., van Daalen, M.P., McCarthy, I.G.: Quantifying the effect of baryon physics on weak lensing tomography. MNRAS **417**, 2020–2035 (2011). arXiv:1105.1075 [astro-ph.CO]
109. Tenneti, A., Mandelbaum, R., Di Matteo, T.: Intrinsic alignments of disk and elliptical galaxies in the MassiveBlack-II and Illustris simulations (2015). ArXiv e-prints. arXiv:1510.07024
110. Amaro-Seoane, P., et al.: Low-frequency gravitational-wave science with eLISA/NGO. Class. Quantum Grav. **29**, 124016 (2012). arXiv:1202.0839 [gr-qc]
111. Amaro-Seoane, P., et al.: eLISA: Astrophysics and cosmology in the millihertz regime. GW Notes **6**, 4–110 (2013). arXiv:1201.3621 [astro-ph.CO]

112. Enoki, M., Inoue, K.T., Nagashima, M., Sugiyama, N.: Gravitational waves from supermassive black hole coalescence in a hierarchical galaxy formation model. ApJ **615**, 19–28 (2004). arXiv:astro-ph/0404389
113. Klein, A., et al.: Science with the space-based interferometer eLISA: Supermassive black hole binaries. Phys. Rev. D **93**, 024003 (2016). arXiv:1511.05581 [gr-qc]
114. Koushiappas, S.M., Zentner, A.R.: Testing Models of Supermassive Black Hole Seed Formation through GravityWaves. ApJ **639**, 7–22 (2006). arXiv:astro-ph/0503511
115. Micic, M., Holley-Bockelmann, K., Sigurdsson, S., Abel, T.: Supermassive black hole growth and merger rates from cosmological N-body simulations. MNRAS **380**, 1533–1540 (2007). arXiv:astro-ph/0703540
116. Sesana, A., Gair, J., Mandel, I., Vecchio, A.: Observing gravitational waves from the first generation of black holes. ApJ **698**, L129–L132 (2009). arXiv:0903.4177 [astro-ph.CO]
117. Wyithe, J.S.B., Loeb, A.: Low-frequency gravitational waves from massive black hole binaries: predictions for LISA and pulsar timing arrays. ApJ **590**, 691–706 (2003). arXiv:astro-ph/0211556
118. Colpi, M., Dotti, M.: Massive Binary Black Holes in the Cosmic Landscape. Adv. Sci. Lett. **4**, 181–203 (2011). arXiv:0906.4339
119. Mayer, L.: Massive black hole binaries in gas-rich galaxy mergers; multiple regimes of orbital decay and interplay with gas inows. Class. Quantum Grav. **30**, 244008 (2013). arXiv:1308.0431
120. Colpi, M.: Massive binary black holes in galactic nuclei and their path to coalescence. Space Sci. Rev. **183**, 189–221 (2014). arXiv:1407.3102
121. Snyder, G.F., et al.: Galaxy morphology and star formation in the Illustris Simulation at z = 0. MNRAS **454**, 1886–1908 (2015). ISSN: 0035-8711. http://adsabs.harvard.edu/abs/2015MNRAS.454.1886S
122. Snyder, G.F., et al.: Diverse structural evolution at $z > 1$ in cosmologically simulated gal axies. MNRAS **451**, 4290–4310 (2015). ISSN:0035-8711. http://adsabs.harvard.edu/abs/2015MNRAS.451.4290S
123. Jonsson, P.: SUNRISE: polychromatic dust radiative transfer. ... MNRAS **372**, 2–20 (2006)
124. Turk, M.J., et al.: yt: A multi-code analysis toolkit for astrophysical simulation data. ApJS **192**, 9 (2011). arXiv:1011.3514 [astro-ph.IM]
125. Pfister, H., et al.: The birth of a supermassive black hole binary. MNRAS **471**, 3646–3656 (2017). arXiv:1706.04010
126. Bowen, D.B., et al.: Quasi-periodic behavior of mini-disks in binary black holes approaching merger. ApJ **853**, L17 (2018). arXiv:1712.05451 [astro-ph.HE]
127. Farris, B.D., Duffell, P., MacFadyen, A.I., Haiman, Z.: Characteristic signatures in the thermal emission from accreting binary black holes. MNRAS **446**, L36–L40 (2015). arXiv:1406.0007 [astro-ph.HE]
128. Lousto, C.O., Zlochower, Y., Campanelli, M.: Modeling the black hole merger of QSO 3C 186. ApJ **841**, L28 (2017). arXiv:1704.00809

The manufacturer's authorised representative in the EU is Springer Nature Customer Service Centre GmbH, Europaplatz 3, 69115 Heidelberg, Germany. If you have any concerns regarding our products, please contact ProductSafety@springernature.com

Printed and bound by CPI Group (UK) Ltd, Croydon, CR0 4YY
12/08/2026
02194517-0003